The Future of Telecommunications

An Assessment of the Role of Competition in UK Policy

MICHAEL E. BEESLEY
and
BRUCE LAIDLAW

IEA

Published by
INSTITUTE OF ECONOMIC AFFAIRS
1989

ATLAS
Sole Distributor for North America
4210 Roberts Road
Fairfax, Virginia 22032, USA
(703) 764-2606

First published in May 1989 by

The Institute of Economic Affairs,
2 Lord North Street,
Westminster, London SW1P 3LB

© THE INSTITUTE OF ECONOMIC AFFAIRS 1989

All rights reserved

ISSN 0073-9103
ISBN 0-255 36220-X

Printed in Great Britain by
Goron Pro-Print Co. Ltd.,
Churchill Industrial Estate, Lancing, W. Sussex

Filmset in 'Berthold' Univers 9 on 11pt Medium

Contents

	page
PREFACE	7
THE AUTHORS	9
ACKNOWLEDGEMENTS	10

ONE:
Introduction — 11
  Avoiding Tough Questions — 12
  Plan of Work — 13

TWO:
The Duopoly Policy and Competition — 15
  A. EVOLUTION OF COMPETITIVE POLICY — 16
    Three Stages of Policy Evolution — 17
    Stage I   Liberalisation of Value-Added Services — 18
    Stage II BT Privatisation — 19
    Stage III The Regulatory Implications — 19
  Box 1: The Littlechild Proposals to Encourage Competition — 21
  B. THE DUOPOLY POLICY — 22
    1. Mercury's Obligation and Plans — 24
    2. Long-Distance Competition — 26
    3. Interconnection — 26
      The Interconnection Determination — 27
    4. Competition in International Services — 29
  C. THE DUOPOLY AND LICENSING — 31
    1. Payphones — 31
    2. Value-Added and Data Networks — 32
    3. Branch Systems — 33
    4. Implications — 33

D.	OFTEL'S PRICE REGULATION	34
	Severe Constraints on Competition	35
	Oftel's View of BT's Profits	38

THREE:
Potential Entry Into Voice Telephony in the UK 41

A.	MERCURY'S IMPACT AND PROSPECTS	42
	1. Connection Cost Barrier	43
	2. Mercury's Non-Competitive Strategy	45

Box 2: Mercury's Revenues 1988-89 45

B.	LARGE PRIVATE USERS	46
	1. Own Account Operation	47
	US Developments in Own Account Operations	47
	2. Motives for Own Account Operation in the UK	49
	Duopoly Policy Reduces Incentives for Own Account Operation	51
C.	RESALE	53
	1. Simple Resale	53
	2. Closed User Groups	56
	3. Network Management	56
	4. Implications	58
D.	A THIRD PTO?	58
	Three Potential Contenders	58
	British Gas Linked to Mercury?	59
E.	LOCAL NETWORKS	60
	1. Broadband Cable	61
	Tough Task for Cable Operators	63
	2. Regional Networks	64
	A Third PTO Option?	66
F.	MOBILE RADIO	67
G.	SATELLITE SERVICES	69
	Potential of Satellites for Private Networks	70

FOUR:
Future Telecoms Policy 73
   A. PRICE CONTROL 73
   B. OWN ACCOUNT OPERATION 74
      Three-pronged Policy Drive 75
   C. MERCURY 76
      Improving Interconnection 76
   D. LOCAL AND REGIONAL ENTRY 78
      New Entrants and BT's Costs 79
      Mobile Radio 80
   E. RESALE 80
      Network Management 81
   F. A STRATEGY FOR REFORM 83
      Improving the Regulatory Process 84
      Improving Flow of Information via Oftel 86
   Box 3: Summary of Recommended Liberalisation Moves in Voice Telephony 84

APPENDICES:
   1. Relationships Between the Regulatory Authorities 87
   2. A Refinement of RP1-X 88

SELECT BIBLIOGRAPHY 92

GLOSSARY 93

LIST OF TABLES
   1. Interconnected Calls: Margins Available to Mercury 29
   2. Potential Competitors in Voice Telephony 42

Preface

The privatisation of British Telecommunications was an historic landmark. It was the first utility and largest trading enterprise to be privatised and propelled the phenomenon of regulation into prominence. BT is now the longest-serving example of a private but regulated industry and provides a set of rich examples of the relationship and conflicts between privatisation, competition and regulation.[1]

A feature of the privatisation of BT is that, in the eyes of many informed commentators, it halted, frustrated and has delayed attempts to promote competition in Britain's telecommunications sector. The liberalisation of value-added network services and the permission and encouragement the Government gave to a new competitor to BT—Mercury Communications Ltd.—were bold steps taken before privatisation. Yet in the lead-up to the sale of 50·2 per cent of BT's equity, Mr Kenneth Baker (then Minister for Information Technology) in a press release made a major statement of government policy—that BT and Mercury were to be the only public telecommunications operators to be permitted to provide basic voice telephony. That is, as part of the strategy of privatisation these two companies would have an exclusive privilege, a legally protected position in the market-place. This 'duopoly policy', as it has become known, is now the major obstacle to the introduction of competition which would benefit the consumer.

Professor Michael Beesley and Bruce Laidlaw are well known experts on the subject of telecommunications and its regulation. Professor Beesley, as the author of a report on the liberalisation of value-added services,[2] and Mr Laidlaw, as an economic adviser in the Department of Trade and Industry during the passage of the Telecommunications Bill, 1984, advised the Government on key aspects of privatisation and liberalisation. Now they review the record. Their conclusion is that not enough has been done to maximise competition and that the basis for further reform is

---

[1] This is the theme of the contributions to *Privatisation & Competition—A Market Prospectus*, London: IEA, 1989.

[2] M. E. Beesley, *Liberalisation of the Use of British Telecommunications' Network*, London: HMSO, 1981.

flawed. The legislative framework in which Oftel (the regulatory body) must operate, does not permit competition to act as the spur for greater efficiency by BT. The danger is that the regulatory route will be adopted which requires the regulator, not the market-place, to deal with high profits, poor services and monopoly abuses. The danger is more regulation and less competition in the future.

The duopoly policy is designed to protect BT and Mercury from competition. Its impact is felt much more widely than reduced pressures on BT, and to a lesser extent Mercury, to be innovative, efficient and responsive to their customers. It infects and limits the growth of other industries, such as cable television and electronic manufacture. It also considerably weakens the case for privatisation by the continuance of artificial legal prohibitions against competition. Indeed, the evidence is lending credence to this view—BT's measured productivity increases are broadly the same today as when it was a nationalised industry.[1]

The authors have set out to tackle the 'tough question' of how to encourage an effective challenge to BT's monopoly position. They provide a detailed and comprehensive analysis of the history and record of telecommunications policy, and the prospects for more competition. Rather than look at the impact of changes in policy on BT, they focus attention on the legal barriers to entry and the opportunities for profit which new entrants would seek. They identify the areas where competition is likely to emerge and how hard it will strike at BT to the benefit of customers. Rather than leave the discussion there, and hence open to the criticism frequently made by bureaucrats that critical assessments are long on analysis but short on solutions, Beesley and Laidlaw make a number of positive recommendations which will move policy along the competition road rather than the regulatory route.

The IEA is an educational trust which does not express a corporate view. Its authors write in their capacity as experts in their chosen fields and are encouraged to carry their analysis to its logical conclusions without regard to administrative and political practicality. Beesley and Laidlaw have done this admirably—they have asked and sought to answer the tough questions, and have provided a set of recommendations which can furnish the basis of a public debate about the future of telecommunications policy which the politicians and bureaucrats will find difficult to ignore.

April 1989 CENTO VELJANOVSKI

---

[1] M. Bishop and J. Kay, *Does Privatisation Work?—Lessons From the UK*, London Business School, 1988.

The Authors

MICHAEL BEESLEY is a founding Professor of Economics at the London Business School. He was Lecturer in Commerce at Birmingham University, then Reader in Economics at the LSE, and for a spell in the 1960s he became the Department of Transport's Chief Economist. He started the Small Business Unit at the LBS. At the other end of the scale, he has advised on company problems of monopoly and restrictive trade practices and on the relationships between nationalised industries and their Ministries. His widely known work in transport economics and telecoms policy has taken him to such countries as Australia, USA, India, Pakistan, Hong Kong, South Korea, Cyprus and many in Europe. His independent economic study of *Liberalisation of the Use of British Telecommunications' Network* was published in April 1981 by HMSO and he has since been very active as an adviser to the Government in telecoms, the deregulation of buses and the privatisation of the water industry. He has been a Visiting Professor at the Universities of Pennsylvania (1959-60), British Columbia (1968), Harvard Business School and Economics Department (1974), Macquarie, Sydney (1979-80). He was appointed CBE in the Birthday Honours List, 1985; and he became Director of the PhD programme in the same year. In 1988 he became a member of the Monopolies and Mergers Commission.

BRUCE LAIDLAW is an economist, specialising in telecommunications policy and regulation. From 1980 to 1984, he was an economic adviser in the Department of Trade and Industry. In that capacity, he was closely involved in the liberalisation of telecommunications and the privatisation of BT. Since 1984, he has practised as an independent consultant, advising governments and telephone companies in many countries. He is a Founder Shareholder of East London Telecommunications, which holds the cable television franchise for the London Boroughs of Tower Hamlets and Newham.

Acknowledgements

We would like to thank the economists with whom we have discussed telecommunications policy questions, particularly Dr Peter Gist, Mr Nick Hartley, Mr Peter Culham and Mr Jeffrey Wheatley. We are also grateful to Mr Michael Dixon and Dr Cento Veljanovski for their contributions to the argument. There may, or may not, be agreement with our conclusions, but the exchanges have been very rewarding for us.

March 1989

M.E.B.
B.L.

ONE

Introduction

When British Telecom (BT) was privatised in 1984, the Government made two important innovations in telecommunications policy. Mercury Communications, which had already been licensed as a rival public network operator, was elevated to the status of second national network and protected by the assurance that no other such competitor would be licensed before November 1990. Secondly, a new regulatory framework was created, including the post of Director General of Telecommunications. The first Director General is Professor Sir Bryan Carsberg.

The policy of having two national networks competing in virtually all aspects of telecommunications was, and remains, unique. Even in the USA, the source of free-market initiatives, local telephone markets are monopolies. It was only to be expected that problems would arise in implementing the UK Government's policy. These have largely fallen to Professor Carsberg and the staff of his office (Oftel) to resolve. His ability to do so has been restricted by the limits on his authority.

The Government's decision to give Mercury time to establish itself as a viable competitor to BT meant no similar entry by other companies into the telecommunications market. From the outset the Government had established a *duopoly policy*. It did, however, set specific time limits to some of the restrictions on competition and linked the price controls on BT to these time limits. Between July 1989 and November 1990, therefore, a series of major decisions about UK telecommunications policy must be taken.

Some have already occurred or are in progress. In July 1988, the Director General announced that he had reached agreement with BT on the first of these decisions. BT is to be subject to a revised price control rule, for four years from August 1989. Other major decisions will include licences to be granted to private networks and the terms on which leased circuits are used ('resale'), as well as whether more competition in running public telephone networks will be allowed. A consultative document on resale was published in February 1989.

Even though these decisions are pending, there is a reluctance in government to consider telecommunications policy as a whole, how the innovations of 1984 have worked out in practice and what to do next. A policy review process was set in train in 1987 when an advisory committee (called the Communications Steering Group) was formed under a senior civil servant, Alistair Macdonald, at the Department of Trade and Industry (DTI). The committee's report, *The Infrastructure for Tomorrow*, was published in December 1988, together with a study by consultants. The report's recommendations for the period after November 1990 are bland. In sharp contrast to the style of the Broadcasting White Paper, published the previous month, the authors of *The Infrastructure for Tomorrow* appear to want to quieten down debate on the key questions.

The report's conclusions are pro-competitive in tone, and to that extent to be welcomed. We agree with their recommendation that any national broadband grid based on optical fibre 'should not be subsidised' (Recommendation 1). Entrepreneurs should indeed be 'as free as possible to use the technologies of their choice' (Recommendation 2). Work on standards should have continued 'high priority' (Recommendation 3). It will be useful, in particular, for the Director General to see how far it is 'feasible' for BT to be made to 'share its ducts, poles and wayleaves' (Recommendation 6). And no-one could rationally oppose a role for government in improving 'user awareness of new communications developments' (Recommendation 7).

Avoiding Tough Questions

But when it comes to tough questions about competitive developments in the future, the Steering Group is ambiguous. Thus, Recommendation 4 says:

> 'BT should continue to be able to play its part in helping to maintain the coherence of the fixed link infrastructure. However, BT should not be allowed to add to its existing market dominance or to establish pre-emptive positions in any new market areas which develop.'[1]

Recommendation 5 says that the Government should promote and introduce competition 'wherever feasible'. Regulation should be based on 'separating the provision and carriage of the services the user wants'. And finally, Recommendation 8 says that the Government should

> 'consider anew the long-term needs of the UK communications infrastructure in several years time, certainly not more than five years'.[2]

---

[1] Communications Steering Group, *The Infrastructure for Tomorrow*, London: HMSO, 1988, p. x.

[2] *Ibid.*, p. xii.

Policy in telecommunications can and should be much more immediately pro-competitive than these words imply. Part of the trouble with the report lies in its method of analysis, particularly in dealing with regulation. The approach so far adopted by the DTI and their consultants is to describe what is technically possible and trends in costs, then to compare likely outcomes under different regulatory policies. Sharply focussed conclusions cannot emerge from this approach. We present a different analysis, based on how profitable entry is likely to be.

The reluctance to debate telecommunications policy stands in sharp contrast to the excitement of interest in information technology generated by the Government in its first term of office. Five years after the Government signalled its intention that telecommunications in the UK should have a competitive future, there is little analysis of what is feasible or desirable. At the centre of the policy problem is how to encourage an effective challenge to British Telecom. This is what Mercury was expected to do, but it has been only a qualified success. Where BT has not had an established market position, the duopoly policy can be fairly said to have been highly successful. The outstanding example is in cellular radio services, where the duopoly between the Vodafone and Cellnet networks started without positions to defend. These duopolists are running a race from the same starting point, of zero production. The success of cellular radio can be measured by the fact that Racal has now been able to sell its telecommunication activities for £1·75 billion.

In other areas, the Government's record since 1980 when it embarked on liberalisation is also good. In addition to cellular, other mobile radio services have been licensed (such as the Band III public networks and Telepoint). The UK equipment manufacturing companies have at last arrested their decline and have begun to rationalise production and marketing. The consumer has a greater choice of equipment than in other European countries. Looking back over the 1980s, it is clear that liberalisation was the appropriate policy response both to the fundamental technological changes affecting the sector and to the failings of BT.

Plan of Work

We do, however, have strong reservations about the Government's telecom policy, in particular for the operation of public telephone networks. This accounts for 75 per cent or more of economic activity in the telecommunications sector.[1] We explain where present policies are having adverse effects and propose alternatives. Our main theme is that more competition is possible and should be permitted immediately. Our analysis does not take the Irish advice to select a new starting point. We build on present realities, not on what might have been done in 1984. In our view, competitive prospects in telecoms depend essentially on the conditions surrounding entry: who could hope to enter the market, make profits and survive, and what regulations would help or hinder them. We identify feasible strategies for encouraging entry and recommend a shift in Oftel's focus towards making entry more effective. We also address the relation between price control and entry. We propose specific reforms in the way in which price control is conducted to realign it with competitive developments. The changes we argue for can mainly be accomplished within the present legal framework. But our proposals also entail some reform of the 1984 Act.

---

[1] We estimate that 1988 output of the UK telecommunications sector, including manufacturing as well as services, was about £11·9 billion, made up as follows:

Public telephone networks: BT	£8·8 billion
Mercury	£0·1 billion
Private telephone networks:	£0·4 billion
Mobile radio services:	£0·5 billion
Data communications:	£0·5 billion
Manufacturing:	£1·6 billion

TWO

The Duopoly Policy and Competition

An appreciation of the evolution of the duopoly policy is essential both to clarify current issues and to consider how far decisions already taken are likely to be sustained or modified. In this section we address the questions:

o How did the duopoly policy come about?

o Of what does it now consist?

o How does the policy affect competition between BT and Mercury?

o What has been its effect on the pace of liberalisation and development of competitive entry?

The duopoly policy was announced on 17 November 1983 by Mr Kenneth Baker, then Minister of State for Industry and Information Technology, in a statement to the House of Commons committee considering the Telecommunications Bill. The key sentence in Mr Baker's statement was:

> 'To avoid uncertainty the Government have now decided to make it clear that we do not intend to license operators other than BT and Mercury Communications Limited to provide the basic telecommunications service of conveying messages over fixed links, whether cable, radio or satellite, both domestically and internationally, during the seven years following this statement.'

'Conveying messages over *fixed links*' defines the function of a telephone network. A telephone network can deliver many services in addition to the voice messages that comprise ordinary telephone calls. Communications from computer to computer, from broadcasting studio to satellite as well as private communications are all caught by the duopoly as stated in 1983. Although Mr Baker's statement went on to indicate that different arrangements would apply in the mobile radio communications field and for cable television, the broad scope of the duopoly has generated a host of difficulties for the regulator. Our discussion touches on these boundary problems, but we treat as the centrepiece of policy concern the market for voice telephony.

Formally, then, the present policy is for a *'fixed links duopoly'*. We refer simply to the duopoly policy because it is now clear that BT and Cable and Wireless (C & W) are accorded special treatment in all aspects of telecommunications policy. Indeed, policy is being developed and applied in a process of continual negotiation in which the Government, Oftel, BT and Mercury are the permanent participants. From this position, the duopolists exercise a strong influence amounting to an effective power of veto over specific competitive developments.

A. EVOLUTION OF COMPETITIVE POLICY

The aim of securing a transition to a competitive environment in the telecommunications sector is characteristic of a wide-ranging change in industrial policy in the UK. The origins of this change are well known. The relatively poor performance of the nationalised industries was attributed to their monopolies in the markets they served, to public ownership and centralised control (particularly of their finances), and to the consequently feeble incentives and disciplines to be efficient for managements and workforces. That is to say, competition in factor markets, for labour and for capital, was at issue, as well as competition in the product market. Telecommunications policy has, even so, focussed almost exclusively on developments in the latter market.

Telecommunications was different from other nationalised industries. Its transformation from civil service into public corporation continued throughout the 1970s. Demand for services was growing much more rapidly than the economy as a whole. In the 1970s, the potential of new technologies to expand the scope of telecommunications services began to be appreciated. To equip the Post Office to cope with the growing importance of its telecommunications activities, they were separated from the postal side. British Telecom was created in 1981 to run these activities. Although major plans for the modernisation of the telephone network were slowly prepared, government control of prices and of borrowing limits prevented investment being expanded in response to the surge of demand. The modernisation programme also fell years behind schedule.

The new Conservative Government of 1979 gave a high priority to opening up the telecommunications market to competition, particularly in innovative services. However, it quickly became apparent that commercial prospects were not good if new firms were confined to innovative services. Competition in *voice telephony* had to be permitted. At the same time, the Government maintained strict controls on borrowing by the nationalised industries, effectively preventing modernisation of the BT network from being accelerated. Attempts to resolve the conflict between these aims led the Government swiftly towards privatising BT. Concern over how a privatised telecommunications company would be constrained to act in the public interest then produced rapid changes in regulation. The duopoly policy was one of the results.

Three Stages of Policy Evolution

The rapid evolution of policy that took place between 1979 and 1983 had three distinct stages. Privatisation first speeded up and then slowed down the pace of liberalisation. In the first stage, which lasted until July 1981, the Government was undecided as to what form competition with BT could or should take. The British Telecommunications Act 1981 gave the Secretary of State the power to license other network operators and service providers but otherwise left intact BT's exclusive privileges. No definite statement of how the Government intended to use its new licensing authority was given at the time.

The debate in Britain on the prospects for competition was conducted largely by reference to US experience. The fundamental issue appeared to be: Was telecommunications a natural monopoly, or could competition be sustained? (What *'natural monopoly'* meant was never precisely formulated, but for most policy-makers it simply meant that a single firm in telecoms would have lower unit costs than would many firms.) The main competitive choice that was posed was between licensing a new long-distance network that would use *digital transmission* technology from the outset (the Mercury project) and removing restrictions on the use made by customers of transmission capacity leased from BT (that is, the resale of capacity issue).

It was widely believed that competition in the USA had developed mainly through the activities of resale operators leasing long-distance lines from AT&T (the long-standing Bell monopoly) and connecting them to local networks. Although this was a mistaken idea (since the principal market entrants in the USA constructed their own facilities and originated as private networks (MCI) or local networks (US Sprint)), resale did seem to offer a rapid means of introducing competition into telephony. Transmission capacity could be leased from BT in a matter of months.

However, BT had inherited a tariff structure in which, for no good commercial or social reasons, leased lines were priced cheaply compared to dialled calls. It was argued that resellers would merely exploit these differences to generate profits at the expense of BT. Since BT was required to make a certain return on its assets, it would be obliged to adjust its tariff by increasing charges for leased lines and, perhaps, for exchange line rentals and local calls.

The discussion on *'tariff rebalancing'* (that is, the alignment of prices to costs) that this initiated and which has continued to the present day has been conducted throughout with remarkably little data on the scale of the profits at risk or of the price adjustments that would be necessary. Rapid price changes for the basic telephone service of unquantified scale were not attractive to a government which attached the highest priority to controlling inflation.

Stage I: Liberalisation of Value-Added Services

The study on *Liberalisation of the Use of British Telecommunications' Network*[1] undertaken by one of the authors and presented to the Government in January 1981, argued as part of the case for competition that resale operations could not be effectively separated from basic or enhanced services, either by regulation or in terms of market developments. This argument provoked an intense debate within the Government, and this debate delayed publication of the study's findings until April 1981 and continued for several months thereafter. Although the Beesley report was influential in the Government's decision to adopt a relatively liberal licensing regime for value-added services involving chiefly data transfer, resale of voice telephony as a basic service was deferred. After protracted debate on exactly how to define them, a general licence for value-added network services (VANS—see Glossary) was finally issued in October 1982.

In his statement of 30 July 1981 welcoming the report and announcing these decisions, Kenneth Baker, then Minister for Information Technology, also said:

> 'a detailed application has been received from Cable and Wireless, BP and Barclays for a licence to provide a business transmission system (project Mercury). The Government are giving active consideration to this proposal and are in principle in favour of such a development'.

A licence was issued to Cable and Wireless in February 1982 to

---

[1] By M. E. Beesley, London: HMSO, 1981.

operate a domestic network, with permission to offer voice services and to connect to BT but limited in scale to 3 per cent or less of BT's revenues from voice services.

Stage II: BT Privatisation

The statement on 30 July 1981 concluded with a reference to BT's investment programme and the problems of financing it. Concern over how BT could modernise its network, which had been building up for some time, initiated the second stage in the evolution of policy towards competition in telecommunications services. During this stage, from July 1981, attempts to exempt BT from constraints on public expenditure were made but failed and a decision taken to privatise the company. This decision was announced in a brief White Paper published in July 1982, simply reproducing a speech in the House of Commons by the Secretary of State.[1] It is outside the scope of this *Research Monograph* to trace the origins of the decision to privatise BT, but it appears to have been the delays in the modernisation programme (particularly the failure of BT to invest even the limited sums agreed by government) that were decisive.

Stage III: The Regulatory Implications

The third stage, from July 1982 to November 1983, was focussed on the implications of privatisation. BT had effectively been a self-regulating nationalised industry with a statutory monopoly. Again, the first point of reference was the USA. It was assumed that a direct control on the return on assets (rate-of-return control) was required, together with a new body to supervise BT's conduct in the market-place. But the prospective difficulties of regulating in private ownership a monopoly as powerful as BT, evidence that US experience with such regulation had not been happy, and the inconsistency between establishing a régime of detailed supervision and the Government's policy of less regulation of the private sector, accelerated the search for a different approach to regulation and for sources of effective competition.

In February 1983, a study by Professor Stephen Littlechild recommended that a 'light' regulatory régime could be achieved if coupled with a major extension of competition.[2] (Littlechild's main recommendations are listed in Box 1.) Central to this approach was

---

[1] Secretary of State for Industry, *The Future of Telecommunications in Britain*, Cmnd. 8610, London: HMSO, 1982.

[2] S. C. Littlechild, *Regulation of British Telecommunications' Profitability*, London: HMSO, 1983.

the idea that price control should be confined to restricting the overall level of prices of services supplied by BT under monopoly conditions (the *RP1-X formula*). The study suggested that several difficulties could be solved simultaneously.

o By setting an overall limit on the rate of increase of tariffs for voice telephony services the need for detailed intervention on prices and profits was removed.

o By avoiding controls on individual prices, tariff rebalancing could continue without resulting in a net increase in the real cost of using the telephone.

o By linking the price control régime to major extensions of competition, substantial incentives were set up for BT to adopt a more commercial approach to the provision of services.

The study was welcomed by the Government and its principal recommendations on price control were implemented, although in modified form.

The strongly pro-competitive tone of policy at this stage did not last. Of the major extensions of competition proposed by Professor Littlechild to balance the 'light rein of regulation' (see Box 1), three, unrestricted resale, international resale and competitive local services were not implemented, three (changes in spectrum allocations, domestic satellites and closed user groups) appear to be proceeding in a very slow and piecemeal way, leaving just two (ending the monopoly of the first instrument used by the consumer to connect to BT, and international interconnection for Mercury) to become features of the Government's telecommunications policy. Cable TV networks were to be allowed to carry public switched voice traffic, but only in conjunction with either BT or Mercury.

A key element in the new strategy was that Mercury would be allowed to develop as a major rival to BT. Mercury was granted permission to provide a full range of international as well as domestic services. Limitations on the scale of its operations were lifted. But it soon became clear that Cable and Wireless and its partners in the Mercury project (BP and Barclays Merchant Bank) were making slow progress in planning a network and finding customers even for its specialised transmission services. It was going to take several years for effective competition in basic services to appear. From this time on, Mercury and its backers became increasingly opposed to further liberalisation.

The main reason for the ebbing of the pro-competition tide and its channeling into encouragement for Mercury was, however, from

> **BOX 1**
>
> ## The Littlechild Proposals to Encourage Competition
>
> o Cable TV networks and private networks owning rights of way should be permitted to compete with BT and Mercury in interactive services (including local telephone services).
>
> o Regulatory restrictions on the quantity of radio spectrum allocated to fixed and mobile communications should be eased.
>
> o A domestic satellite network should be licensed in competition with BT and Mercury.
>
> o Shared use of private networks among closed user groups should be permitted.
>
> o BT's monopoly of the first telephone instrument should be ended.
>
> o Unrestricted resale of capacity leased from BT should be permitted.
>
> o Mercury should have access to public networks overseas and users be permitted to resell capacity on leased international circuits.
>
> *Source*: S. C. Littlechild, *Regulation of British Telecommunications' Profitability*, Department of Industry, London: HMSO, 1983, para. 14.8.

July 1982, the pressure created by privatisation. Privatisation of BT was a transfer of ownership on a larger scale than had been attempted before in UK financial markets. Before the event, the Government and its advisers were uncertain that it could be done at all. In this atmosphere, it was considered prudent not to disturb the markets in which BT operated by further extensions of competition or radical changes in the regulations BT had operated under. These might put at risk BT's profits, if successful, or lead to more onerous regulation, if not. So far as new networks were concerned, the sale of shares in BT and the slow growth of Mercury seemed, to City opinion as relayed to the Government, to preclude any other market entrant from raising substantial capital in financial markets.

BT was to be privatised with its management and internal procedures virtually unchanged. Nevertheless, the possibility of restructuring BT into independent regional companies with separate national and international networks was widely canvassed. BT's management believed that it would take some years to place the company on a fully commercial footing. Although running a company with 100 per cent of the market, the management felt vulnerable to competition while internal reorganisation was proceeding and this feeling was communicated to financial markets. That separation was physically feasible was not in doubt (there was after all the precedent of Hull, a profitable independent local network). However, BT's management in the regions was judged to be weak and, it was believed, could not cope with immediate exposure to commercial realities. In any event, creation of accounts for separate regional networks would take too long for the Government's timetable.

B. THE DUOPOLY POLICY

The duopoly policy was announced by the Government in a wide-ranging statement covering future competition policy in telecommunications.[1] The statement was made following the release of the first draft of the proposed BT licence and was intended to provide the background to it. In the statement, reasons given for restricting entry into telecommunications were:

o To allow Mercury to reap rewards from the 'considerable investment' in a national public telecommunications network that will be required of it under a licence 'very similar to that proposed for British Telecom'.

o To give BT time to adjust its organisation and services towards the competitive future.

o To husband scarce radio frequency resources.

o To prevent disruption of the environment by having a number of operators installing cables in the same area.

Notably absent is any reference to the many doubts that had been expressed, particularly in the House of Commons in the previous session, that a rapid introduction of competition would place at risk BT's ability to provide services to residential and especially rural customers. This issue had been resolved by placing in the Bill a statutory duty on the Secretary of State and the Director General of Telecommunications to ensure that such services could be financed.

---

[1] K. Baker, 'Statement to the Standing Committee on the Telecommunications Bill', 17 November 1983.

The conclusions announced by the Government in November 1983 did not accord well with the earlier debate on the scope for competition. That debate had been about how far voice telephony services would be open to competitive entry and whether entry could arise from new facilities or from removing restrictions on the use of existing facilities (resale). The statement made no direct reference to voice telephony and deferred decisions on resale. In consequence, the positions of several classes of potential entrants were not explicitly considered. These loose ends required further statements and, even so, have created continuing difficulties with licensing.

At the time, the main point in dispute was the duration of the duopoly. The Government's difficulty was that, having raised the status of Mercury to be the national challenger to BT's monopoly, it became dependent on the willingness of Mercury's owners to invest on a sufficient scale to make the policy work. Seven years emerged as the minimum period Cable and Wireless would accept in return for taking on the risks of entry. This was longer than was thought necessary to give BT time to adjust to privatisation. Similarly, Mercury's fears about the impact on its prospects of the resale of leased lines were greater than BT's. The duopoly policy therefore promised less competitive pressure on BT than BT's management would have been willing, if pushed, to tolerate.

A subsequent statement on resale and private networks, made by written answer to a Parliamentary Question on 20 July 1984, addressed some of the issues left undecided by the 17 November statement. By this time, the tone of the Government's policy on further competition was markedly less enthusiastic. There had been a wide-ranging agreement between the DTI, BT and Mercury on the content of licences and on the terms of interconnection between the two networks (Heads of Agreement on interconnection were signed on 14 June 1984). The statement confirmed that resale as a long-distance voice service interconnecting with BT's local networks (termed 'simple resale') would not be permitted before July 1989. Conditions to this effect were to be included in all licences. The rules for private networks interconnecting with *public networks* (the typical *PABX networks* run by businesses) would be essentially a codification of BT's existing rules, with a limited exemption for specialised business networks. In setting out the criterion for licensing such networks, the principle of preventing revenue diversion from BT was endorsed for the first time (the actual phrase used was diversion of 'traffic revenue', the precise meaning of which is not clear).

This further restriction of competitive opportunities may be attributed to the pressures of the privatisation timetable. The small team of government officials charged with devising the new set of rules simply did not have the time or the resources to devise a policy on interconnection between the BT and Mercury networks. It was left to the two companies to negotiate. Their outline agreement in June 1984 was crucial in securing the consent of BT's management to the Government's package of privatisation and regulation. In the event, this agreement was quickly repudiated by Mercury.

1. Mercury's Obligations and Plans

The duopoly may be characterised as competition between two companies with permission to engage in the profitable business of voice telephony in return for accepting service obligations of a non-commercial character. Of these, the main obligation on each firm was to operate a national network, that is, to provide a voice telephony service uniform in quality and price throughout the country. However, Mercury was in no position to commit itself to carry out an obligation to provide services nationally until it had built up its network and customer base. The period until 1990 would perhaps give the necessary time. But the incentive Mercury would have for voluntarily undertaking extensions of its network into rural and remote areas was not established. The Government may have had it in mind that failure by Mercury to do so could in principle have been met by additional licence obligations. Nevertheless, it is barely conceivable that additional obligations would be written into the Mercury licence in 1990 just when it was possible that the market was to be opened up to competition from others. Given the problem of market development faced by Mercury and the intention to broaden the scope for competition at a later date, the duopoly policy, in so far as it looked forward to a second national network alongside a 'universal' one, was very difficult to realise.

The Mercury project, before privatisation of BT was announced, involved the construction of new facilities over which a variety of services could be provided to business users. The technologies to be employed were to be *digital*, rather than *analogue*—as was BT's network at that time. The costs of construction were to be reduced by laying high-capacity optical fibre cables along railway lines or by using microwave radio point-to-point transmitters. At the outset the services to be provided over these facilities were not clear. It took some time for the equity partners in the project, Cable and

Wireless (40 per cent), British Petroleum (40 per cent) and Barclays Merchant Bank (20 per cent), to prepare and agree plans.

Two main markets were envisaged, long-distance transmission between major business centres and users concentrated in the City of London. Mercury's principal investment was in an optical fibre cable laid in a figure-of-eight between London, Bristol, Birmingham, Manchester and Leeds. In London, a local duct network was purchased from the defunct London Hydraulic Power Company (LHP) and converted to carry telephone cables. Among the many options for the new company, two alternative strategies were widely debated. It was possible for Mercury to pursue a strategy of being the 'second source' of communication links for major business users, or it could develop a full-scale network. The former strategy would minimise the scale of investment and, therefore, the risk to the project's backers; the latter would open up the main market for switched voice services. The first licence issued to Mercury in February 1982 did not preclude entry into switched voice services (international services were excluded and Mercury was required to offer digital transmission); the decision was left to the partners in the Mercury project. Interconnection with the BT network was envisaged as the means to collect and deliver calls. An agreement to this effect was made between the two companies later in 1982. However, Mercury did not offer switched voice services for a further four years.

The original licensing of Mercury produced a strong reaction from BT. In particular, BT accelerated plans to provide digital transmission services, by installing a *microwave* 'overlay' network in central London and planning a nationwide network for its *Kilostream* and *Megastream digital private circuits*. BT also began a staged restructuring of its tariff for analogue leased lines. This increased charges substantially, particularly for short distances. These changes may have been made primarily in the expectation that resale might be permitted in some form, but the effective exclusion of Mercury from analogue transmission services and the high cost of short-distance microwave links must have been important factors.

In May 1982, BT sharply reduced tariffs for long-distance calls between major cities, including those on the railway routes Mercury was planning to use. Considered solely as a response to Mercury, these tariff reductions appeared to be both premature and an over-reaction, but BT's increases in tariffs the previous year (November 1981) were believed to have been excessive. Selective reductions in long-distance tariffs were therefore part of the tariff rebalancing programme—and also an early move towards 'de-averaging' tariffs.

2. Long-Distance Competition

Despite these tariff changes, the margins on long-distance calls remained attractive. Undoubtedly, the prospective entry of Mercury into this market was the strongest single influence on the Government's decision to defer permission for resale of leased lines. Mercury would provide the competition to BT; by using alternative facilities Mercury could offer users a better choice than could resale operators dependent on price differences between services using the same transmission facilities. Mercury's own perception of its prospects was that a prohibition on resale was not necessary (nor was it undesirable). Its entry into switched voice services required, above all, improved terms of interconnection so that it could offer itself to users as an alternative network to BT, while handing back to BT traffic it was unable to deliver, or which it was unprofitable to handle. Interconnection developed into the central regulatory issue.

3. Interconnection

The principle that Mercury should be able to connect its network to BT's was accepted at an early date, simply because it was inconceivable that interconnection should be denied to a public network. The debate centred on the terms on which interconnection could be achieved. There were two basic issues:

o   At what point within the BT network would interconnection take place?

o   What charges would be appropriate for interconnected calls?

For Mercury, what was at stake was the scale of investment it would be required to make to reach customers and what margins would be available on each service. Access to the BT network at a high level would enable it to reduce to a minimum the number of exchanges to be built (in the limit, only one). If charges were levied that were not substantially lower than BT's tariff, there would be no margin for Mercury. In addition to its concern to constrain Mercury's prospects, BT had a legitimate concern that giving Mercury preferential terms for access to its customers would leave it only with low-margin business. These issues could not at first be resolved by the Government. The licences therefore set out the principles of interconnection and the basis on which rules could be determined by the Director General should BT and Mercury fail to agree.

So, largely at the insistence of BT, the two companies reached outline agreement on interconnection before BT was privatised. Once this pressure was removed, disagreements came to the surface again and Mercury appealed to the Director General. Disputes over interconnection have continued ever since.

The Interconnection Determination

The Interconnection Determination made by the Director General in October 1985 was attractive enough for Mercury to commence switched voice services the following May. Mercury was granted direct access to BT's trunk exchanges, so that it did not need to build local exchanges itself. Mercury was required to pay the full cost of interconnection. The schedule of payments for interconnecting calls allowed a substantial discount on BT's tariff for local and long-distance calls, though not for international ones. However, the Determination left many issues in doubt and sufficient remaining disabilities on Mercury to constrain the pace of its development as a rival to BT.

In terms of the quality of interconnection, the principal difficulties have arisen over the need to connect Mercury's digital network to BT's still largely analogue network, and BT's tactics. The Determination provided for BT to undertake the work of connecting the two networks, but for Mercury to pay. BT refused to tell Mercury where it had installed digital exchanges, or when it planned to replace analogue exchanges by digital. As BT continues to run analogue trunk exchanges in parallel with digital exchanges, BT has been able to connect Mercury mainly to old exchanges. This tactic has induced serious problems of quality for Mercury's customers. Calls that are routed over BT's network take noticeably longer to be connected and have a higher than average failure rate. The tactic has also increased the expense of interconnection.

Difficulties with interconnection have seriously hindered Mercury's expansion. Some 80 per cent of the connections requested by Mercury to date have been delayed by BT beyond the time allowed for them. BT has also adopted a very restrictive attitude towards the risk that calls carried by it may not be audible, effectively preventing half the telephone customers in the country from accessing Mercury. Difficulties have also arisen over all the ancillary services involved in running a telephone service. It would benefit both users and the duopolists themselves if they co-operated on the provision of emergency services, directory enquiry services and operator fault enquiry services. Each service has, however, been the subject of running disputes. With ancillary services for which a charge is made, such as the 0898 services, it is naturally not in BT's commercial interest to co-operate in providing Mercury with the means of access. BT has so far been able to exploit the vagueness in the Interconnection Determination to prevent Mercury from offering them.

The Determination provided an arbitration procedure for resolving some of these difficulties. This procedure has so far worked slowly and with little effect. The most intractable problem referred to it is BT's charges to Mercury for interconnection.

The Director General in his Annual Report for 1986 said that the arrangements he devised for charging for interconnection were based on costs, but did not elaborate. Three types of cost are, presumably, involved. First, there are the initial costs of interconnection. They are estimated by BT and charged to Mercury as the work requested by Mercury is completed. These could not be allowed for explicitly in the payments per call set out in the Determination. Second are the attributable costs associated with making particular calls, and third are the costs associated with expansion of the total network as call volumes increase. Only the first of these types of costs have been referred for arbitration. As argued later, we do not think that these costs can be estimated without very debatable and complex assumptions, beyond present capabilities. The result is that the interconnection charges must in practice have been fixed with reference to what the traffic—in this case Mercury—would bear. This is a judgement that Professor Carsberg can, and did, make. But it is hard to see the grounds on which an arbitrator can logically intervene.

The important point for competitive entry is that the payments should provide a sufficient incentive for Mercury. At 1985/86 tariffs, the schedule of payments provided an average margin for Mercury of about 50 per cent on BT's call charges, with a lower margin on local calls than on long distance. This appeared to confirm that Mercury's main market opportunity lay in long-distance 'bypass' of BT. Indeed, Mercury's early tariffs for switched voice services discouraged the making of local calls over the network. On long-distance calls, Mercury's tariff offered an average discount of around 17 per cent compared with BT. After BT's response and Mercury's subsequent revision, the discount appears to have widened, mainly because Mercury did not increase its charges for local calls (Table 1). This, plus the higher minimum charge imposed by BT (4·4p per call compared to 3p by Mercury),

Table 1:
Interconnected Calls: Margins Available to Mercury
(Margins as a percentage of the BT tariff)

		Call duration in seconds			
		15	60	200	300
		%	%	%	%
2100 service					
Local call	C	95	77	25	–
	S	91	77	50	43
	P	90	48	57	48
Short national	C	95	77	63	62
call	S	91	77	75	74
	P	90	83	78	78
Long national	C	95	77	81	77
call	S	91	77	79	77
	P	90	83	81	81
2200/2300 service					
Short national	C	91	55	25	24
call	S	90	51	54	46
	P	77	63	54	54
Long national	C	91	55	63	55
call	S	90	51	54	51
	P	77	63	59	60

C = Cheap rate (Economy)   S = Standard rate   P = Peak rate (Prime)

Notes: BT tariff as at 1 January 1989; payments for interconnected calls derived from the Interconnection Determination, October 1985.
Mercury is assumed to hand over calls to BT for final delivery at Level 3J.

means that Mercury is now offering most directly connected customers (2100 service) larger cash savings on local calls than on long distance. For indirectly connected customers (2200 and 2300 services), Mercury does not accept local calls.

4. Competition in International Services

Competition in international voice telephony services has developed along different lines. In discussions following the November 1983 statement, it was agreed that BT and Mercury would conform to common accounting arrangements for international services. The licences granted to them provided for these arrangements to be set by negotiation or, failing agreement, to be determined by the Director General. Although not explicitly stated at the time, this amounted, in the expectation of the parties, to an agreement in principle to share the international market. Mercury's entry into international services therefore depended on its being able to secure operating agreements with overseas network operators. Agreements were secured early on with North American operators, and with overseas networks operated by Cable and Wireless.

In accordance with a market-sharing rather than a competitive policy, Mercury was not given discounted access to BT's international network under the Director General's 1985 Determination. BT and Mercury did not reach agreement on international accounting arrangements and the Director in due course determined a Code of Practice. The Code of Practice provides that BT and Mercury shall share information about agreements reached with overseas operators, thereby giving Mercury access to BT's agreements. Agreements should conform to existing international rules concerning the division of costs of providing international services (these allocate half the notional cost of carrying both incoming and outgoing calls to each operator—termed 'parallel accounting'). Traffic should not be routed through third countries in such a way as to circumvent the rules on parallel accounting.

The Code of Practice is inevitably lopsided. It constrains what Mercury can do to break into the international market, but the UK regulatory authorities can do little directly to help Mercury secure agreements with overseas networks. The Code of Practice does permit Mercury to depart from parallel accounting if unable to secure an arrangement under which traffic incoming to the UK is allocated between it and BT in the same proportion as traffic outgoing from the UK (an arrangement termed 'proportional return'). The main problem for Mercury, however, has been the reluctance of many countries, for example, France and Germany, to sign any agreement for switched voice service.

Mercury has been able to make some use of this Code, having established PSTN (see Glossary) connections to 12 countries. There are many reasons why its progress has not been faster. At least until 1987, there was concern abroad that Mercury might not survive a change of government in the UK. There was a continuing dislike by network operators of competition, even if not aimed directly at themselves. Mercury was also unable to offer a significant incentive to network operators for interconnection. The lack of incentive, given parallel accounting, arises from Mercury's very low domestic market share and its reliance on the same international facilities as BT to reach overseas destinations. Cable and Wireless has ambitious plans to install new undersea cable connections to major overseas markets but these will take some time to come into service. Mercury did not receive the support it needed to persuade foreign administrations to admit it. Recognising these problems, in July 1988, the Director General reduced the level of interconnection payments on routes to countries where Mercury has no operating agreement. This change ensured that Mercury's international services would be profitable, but did not intensify competition.

C. THE DUOPOLY AND LICENSING

Since the duopoly policy was announced, further entry into the voice telephony market has been severely restricted. The principal opportunity was through the operation of local broadband cable networks. Each franchise awarded carries with it a PTO licence, though voice telephony services must be provided in conjunction with BT or Mercury. (As BT already operates extensive local networks, its interest in broadband cable could be only on terms that precluded competition. Franchisees have Hobson's choice— deal with Mercury or leave telephony out of their plans. Mercury, for reasons we explore in more detail later, has not been willing to deal with many franchisees.) In any event, the difficulties experienced by cable television companies in raising finance has precluded their entry into voice telephony.

At the same time, policy has not been static. There has been movement on three fronts, in each case essentially accommodations of licences to practical considerations rather than changes of principle:

o Permission for Mercury to offer a public payphone service and for private users to purchase payphones outright rather than rent from BT.

o A substantial revision of the rules concerning the operation of *value-added and data networks*.

o Minor modifications of the licence for interconnected private networks (branch systems) to bring it into line with that for value-added and data networks and to change some impractical provisions.

1. Payphones

In 1984, payphones were excluded from the duopoly policy and from the liberalisation of the supply of apparatus. Only BT could operate a public payphone service. Also, customers who wanted to have a coin-operated telephone (a private payphone) were obliged to rent the apparatus from BT simply because the DTI and BT agreed in 1984 not to approve any payphones for connection to public networks for at least two years. The expectation at that time was that, without resorting to closure of many existing kiosks, it would take BT some years to put its payphone service on a proper commercial footing. Since privatisation, however, BT's accounting losses have continued while the quality of the payphone service declined. Mercury successfully applied for permission to operate a public payphone service of its own, unveiling the first kiosks at London railway stations in July 1988. To date, about 100 Mercury payphones have been installed. Applications from other companies to run independent public payphone networks are still pending. Oftel has now issued a specification for approving payphone apparatus, which may enable some private payphone services to become established in 1989.

2. Value-Added and Data Networks

The principal policy reason for revising the 1982 *VANS* licence was to give effect to a decision by the Government to prevent BT and IBM from starting a joint venture to provide data services. The joint venture, termed JOVE, proposed exactly the kind of innovative data services that the Government wanted to encourage, but on terms that aroused fears that the market would be monopolised. For the purpose of our discussion, there were two major issues:

o How should voice and non-voice markets be separated in regulatory terms?
o How should the provision of data services by BT and Mercury be regulated?

The 1982 licence achieved the separation of voice from non-voice by requiring that every message passed across a value-added network be acted upon in one of three defined ways (see Glossary). Ordinary voice telephony was thereby precluded. However, many forms of data communication were also precluded, including those proposed by JOVE. The revised licence issued in 1987 permitted voice ('live speech') if a subsidiary element of a data service, or if a premium over the tariff for the equivalent public switched telephone network service was charged. The wording of the relevant licence condition, Condition 2, also prohibited resale of leased circuits for voice telephony. The licence also introduced conditions intended to curb anti-competitive practices. These rules were applied to BT and Mercury as well.

3. Branch Systems

American evidence indicates that major users operating networks on their own account are a significant source of external pressure on the *PTOs* as well as being themselves potential entrants into public network operation. The duopoly policy implied that this would not be permitted in the UK before 1990. As restated in the July 1984 statement, private networks would also not be permitted to divert 'traffic revenue' from public networks. That is to say, the scope for self-provision by private users would continue to be limited. The general licence issued under the 1984 Act for interconnected private networks, the *Branch Systems General Licence* (BSGL), prevented market entry or extensive self-provision by:

o prohibiting the connection to public networks and other private networks of privately-constructed facilities extending beyond a single building,

o limiting the scale and configuration of leased circuits comprised in private networks;

o prohibiting the carriage of third-party traffic except in very limited circumstances.

The formulation of these restrictions in the licence caused problems when applied to networks already in existence. Many companies found that they could not do under the licence things that they had previously been permitted to do by BT and to which there was no objection under the duopoly policy. These companies have usually been granted temporary individual licences. The 1987 revision of the BSGL effectively 'grandfathered' several of these anomalies. It also brought the licence into line with the new value-added and data services licence. But the 1987 revision did not address the competitive boundary between public and private networks. In consequence, several networks providing voice telephony services to third parties remain licensed individually outside the BSGL, and some potential entrants, such as British Rail, continue to operate under temporary licences.

4. Implications

These licensing issues illustrate in different ways what is the most striking aspect of the working of the duopoly today, namely, the dependence of the Government, and of Oftel which increasingly acts as its agent on licensing matters, on the co-operation and consent of BT and, to a lesser degree, of Mercury. In part, *this is the inevitable consequence of regulation being undertaken by govern-*

ment officials—they rely for information and for practical guidance on the firms they regulate. By granting exclusive rights to the two national network operators in unspecific terms, the Government has been obliged thereafter to refer licensing proposals to them prior to a public announcement for assurance that they did not represent a breach of the duopoly. It could be argued that referral was necessary in any event since additional competition might put at risk the services BT and Mercury are obliged to provide, but this in present circumstances would be merely a formality.

The duopoly appears to be interpreted, by government and the general public alike, as meaning that no fundamental change in licensing or regulation of any kind will occur until 1990. In that sense, the precise words used in 1983 matter less than the spirit of the policy that stability is necessary while the great changes wrought by privatisation are absorbed. It is this spirit of stability that is the most powerful deterrent to market entry, since potential entrants are deterred from seeking out loopholes in, or exceptions to, the duopoly policy. The general expectation is that the regulatory authorities would somehow prevent them being exploited. Thus one major avenue of development in telecoms competition which the USA experienced is missing. There, competition grew through challenge to regulatory decisions in the courts.

D. OFTEL'S PRICE REGULATION

We now examine what Oftel has done since 1984, concentrating on the regulation of prices. Oftel's key role in implementing interconnection between BT and Mercury has already been discussed. The dual regulatory structure of licensing by the Secretary of State and policing the activities of licensees by the Director General has in general achieved its objectives. With some modifications, this duality has been reflected in the regulatory arrangements for subsequent privatisations. In our view, however, the arrangement has weaknesses for telecommunications where there are manifold opportunities for new firms to enter the market. These weaknesses have led Oftel away from its original remit and, despite the expression of good intentions, are likely to result in unnecessary delay in the development of effective competition.

Before 1984, BT had itself been the effective regulator; so this function had to be divested. The UK did not have a tradition of regulating private sector monopolies that was adequate to the task. So much was recognised and was the basic reason that Oftel was set up. The new regulatory authority then had to work out for itself how it was to operate within the constraints of the 1984 Act. Oftel's economists, Nick Hartley and Peter Culham, have recently written: 'Oftel's fundamental objective is to achieve productive and allocative efficiency in the telecommunication industry'.[1] This describes the classic objective of a regulator of a private monopoly. But they add:

> 'the new element in Oftel's task is that, so far as this is possible, it also tries to encourage competition. Where entry is possible Oftel tries to ensure that effective competition develops'.

Severe Constraints on Competition

The constraints on Oftel's pursuit of effective competition are severe. Lacking licensing powers, the Director General cannot increase the number of network operators. He is, therefore, not able to challenge the duopoly policy. More generally, he is inhibited from calling in question any part of the 'carefully balanced package' agreed between BT, Cable and Wireless and the Government in 1984. On specific matters concerning the enforcement of BT's licence conditions, or their effectiveness as currently written, he is constrained by BT's right of appeal to the Monopolies and Mergers Commission (MMC), to the courts or the Secretary of State (the complexity of the structure within which Oftel works is described in Appendix 2). In order to preserve his central role, the Director General is drawn into negotiation with BT to secure its consent to being regulated.

Unable to react to inefficiency or the abuse of monopoly by extending competition, how can the Director General respond to complaints from customers about BT? He can only try to tighten regulatory controls. This is what has happened with the price controls on BT.[2] In July 1988, the Director General announced that he had agreed with BT an extension of the RPI-X price control formula from July 1989 for four years. The value of 'X' will rise from 3 to 4·5 and the control will be extended to include connection charges.

A separate price control would apply to *analogue private circuits*, but the details have not yet been announced. As a result of these

---

[1] N. Hartley and P. Culham, 'Telecommunications prices under monopoly and competition', *Oxford Review of Economic Policy*, Vol. 4, No. 2, 1988, p. 3.

[2] The main price control on BT is the RPI-X formula. This provides that the prices charged by BT for the main elements of the voice telephony service (rentals plus inland call charges) should not increase in any year by a percentage that is more than three percentage points below the rate of inflation, as measured by the retail prices index. The control was set in 1984 for five years.

changes, the proportion of BT's revenues subject to price control will increase from about 50 to 57 per cent.

The consultative document on price controls issued by Oftel in January 1988 described two objectives for regulation:

> 'First, regulators should, as far as possible, seek to mimic the pressures of a competitive market ... second, the regulatory structure should be widely accepted as fair ... .'[1]

Fairness depends on which interests are being taken into account; we do not pursue it here. The objective of mimicking the market does, however, raise several problems. It is possible to make assumptions about the terms on which entry could occur and then derive propositions about prices and outputs from these. Alternatively, the blocks on entry can be examined to predict what entry is likely, leaving aside predictions about prices and outputs in the industry as a whole.

Oftel's economic arguments and its practice diverge. Nearly all its analysis adopts the first approach. We think this is of very limited use. To mimic the results of competition involves a great deal of information and very sophisticated models. Oftel does not have the necessary information to make correct interventions. On its record to date, it is not even trying to collect the relevant information. Most importantly, the prices set after competition has arrived may well be quite different from those which Oftel (or anyone else) would arrive at by contemplating the demand and cost conditions now and prospectively available to BT. The same is true of costs. Entry, or the prospect of entry, is the most likely cause of BT achieving a reduction of cost levels not now attained or even believed to be attainable.

To judge BT's prices, and to take relevant action, the prices must be 'unbundled'. That is, each element of the use of the telephone which consumers see as distinct and of which consumption may vary when prices are changed should be paid for separately. Only if prices are unbundled can prices discipline the behaviour of both consumers and producers and the regulator obtain good information about costs. Unbundling of BT's telephone services was predicted to be a beneficial consequence of liberalisation; the price controls were designed to permit it to happen without regulatory constraints. Yet it has scarcely begun. For example, directories are provided 'free'—that is, the same directories are supplied to all customers and charged for in the rental. If users had to pay for

---

[1] Director General of Telecommunications, *The Regulation of British Telecom's Prices*, London: Oftel, January 1988.

directories, but could lower their rental payment by having a less frequent updating of the directory, many might choose that alternative. Others might buy additional directories. As a result, Oftel (and BT) would obtain for the first time good information on the demand for, and cost of supplying, directories.

Unbundling has not occurred for two quite distinct reasons. For BT, unbundling its tariff for telephone service is undesirable because it would increase the opportunities for competition and reduce its room for manoeuvre in responding to market entrants. For Oftel, there is public pressure to contend with. If unbundling was pursued, it would, for example, probably entail BT charging for directory enquiries. There is little public confidence that unbundling could be done without a net increase in telephone charges.

If Oftel is to be an efficient regulator it must identify the cost attached to each soparable element of the telephone service. This also has not been done. Oftel seems to subscribe to the consensus view that a large part of the costs of telephone service cannot be attributed to any particular service, but are incurred in common in running a network. The importance of *common costs* is probably exaggerated, partly because the difficult task of identifying attributable costs has not been undertaken rigorously. However, *the dismal truth is that no one, not even BT's management, has much idea about costs*. Moreover, even if they had full information about the impact of varying service levels on costs, getting at correct prices would involve taking a view on where margins should be taken to cover common costs. Although Oftel's economists have given considerable thought to this issue, the Director General has not lent their calculations much weight.[1]

We may be sure that Professor Carsberg and his colleagues are aware of the practical problems of pursuing their regulatory objectives. When obliged to make a decision about the renewal of price controls, the Director General appears to have placed very little reliance on theoretical considerations about efficient markets. As with other matters, the outcome can be most satisfactorily represented as a bargain. With the slow growth of Mercury and the Government's continued commitment to the duopoly, continuation and some tightening of price controls were inevitable. Given the rate of increase in BT's profits, a modest increase in X

---

[1] The issue arose in the review of price controls. Oftel's economists proposed a pricing scheme that would cover attributable costs of each service plus a mark-up inversely proportional to demand elasticities. This scheme is known to economists as a Ramsey pricing rule, and was referred to by the Director General as 'economic' pricing. He did not endorse this approach.

announced well in advance would not be too unsettling for shareholders.

With the exception of the residential rental, individual prices are not directly affected by price controls. Yet for Mercury and other potential entrants, the details of BT's individual prices are as important as the aggregate level. Since 1982, BT has sought to 'rebalance' its telephone tariff. As Mercury was perceived as a competitor on long-distance services, BT moved at an early date to lower trunk charges. Subsequently, rental and local call charges were increased.

Oftel's response to the rebalancing moves by BT has been focussed almost exclusively on the question whether the new prices are reasonably related to costs, rather than their impact on the prospects of competitors. In January 1988, the Director General expressed the view that local call charges should not rise further relative to long-distance charges. Rental charges were still below cost but should remain so pending further discussions. In response, BT undertook to leave prices unchanged until 1989. We are sceptical that Oftel could reach firm conclusions on the relationship of BT's prices to underlying costs. However, it is clear that the speed of rebalancing was giving rise to a substantial volume of complaints from customers and placing at risk Mercury's move into profitability. These were sufficient reasons to negotiate a pause with BT.

In summary, then, the debate about price controls on BT has been conducted mainly in terms of the relationship of prices to costs, but this relationship is and is likely to remain obscure. Decisions about price controls appear not to have depended in practice upon detailed or accurate cost information. By extending the scope of price controls, the Government's original policy of distinguishing monopoly services, which had to be regulated, from the rest, to be free from inhibitions on making profit, has been implicitly denied.

Oftel's View of BT's Profits

When revising the level of price control, Oftel had to take a view about profits. No attempt has been made to establish a correct profit constraint for those services subject to price control. Discussion has been focussed on the reasonable rate of return for BT *as a whole*. The standard of what is reasonable is to be found in BT's required cost of capital as determined by financial markets. Yet the controlled sector of services will, because subject to less uncertainty, have a lower cost of capital than the rest of BT. In fixing price controls, Oftel in effect looks to BT to make a higher rate of return than would be appropriate for the controlled sector. On the other hand, Oftel regards BT's prospectively higher returns from non-controlled activities, such as international services, to be legitimate sources for reducing prices in the controlled sector. In such ways, the intended distinction between the controlled and competitive sectors is being eroded, implying cross-subsidies of uncertain size.

Of course, Oftel is right to be concerned about the trend of BT's costs. When BT was privatised, it was expected by the Government and by BT's management that network modernisation, more flexible working methods and reductions in the labour force would produce substantial savings in running costs. There was room for doubt about how fast these savings might emerge, but it was expected that within the five-year period of the RPI-3 price control, sufficient would be achieved to justify the policy. The challenge from Mercury was to be an important ingredient.

British Telecom's productivity record cannot be discerned from the limited information in the public domain. It is likely, however, that the company's productivity record has not improved significantly since privatisation. As a nationalised industry, BT had a formal target between 1978 and 1983 of reducing its real unit costs[1] by 5 per cent a year. In practice, it never met this target, averaging only 2·8 per cent a year. Since privatisation, BT has refused to publish either the volume or the tariff indices which would enable its real unit cost performance to be measured. Oftel has taken an agnostic view of how productivity should be measured and so has not insisted that the missing indices be reinstated. In the 1988 Consultative Document on price controls, it stated that 'the likely level of productivity increase seems to lie in the range of 2-3 per cent a year'. We estimate that real unit cost reduction since privatisation has averaged 2·3 per cent a year.[2]

---

[1] Real unit costs are a measure of total factor productivity. Formally, the latter is calculated as total operating costs (deflated by the retail prices index) divided by turnover (deflated by the price index for BT's services).

[2] Over the 10 years to 1988, BT's total operating costs increased at 4·5% per annum, with the rate of increase since 1984 the same as it was before. Between 1984 and 1988, BT's turnover has increased at 10·5% while inflation has averaged 4·5% a year. If BT's prices had been constant in real terms, the growth in volume would have been 5·7% a year. In practice, BT's prices have fallen slightly in real terms at a rate of 1-1·5% a year. The vagueness on prices arises because it is necessary for this calculation to include unregulated as well as regulated prices. Allowing for the price trend, the volume increase has been about 7% a year. Dividing the real rate of increase in total operating costs (4·5%) by the rate of increase in volume (7%) produces a rate of fall of real unit costs of 2·3% between 1984 and 1988.

It is clear that BT has great unrealised potential to improve its costs. The scope is largest in respect of labour costs, where international comparisons indicate that BT has over 75,000 more staff than is warranted by the size of the network. Most of these surplus staff are engaged in local network operations, so far shielded from competition. At privatisation, BT management set itself the target of reducing the labour force employed in the network by 5,000 a year. This modest target was not met. Recently, BT has argued that staff numbers must be maintained while modernisation takes place and improvements in the quality of service offered to customers are carried out.

THREE

Potential Entry Into Voice Telephony in the UK

Increasing competition in telecoms, as in many other industries, depends on challenge to the major incumbent. In the UK, that challenge is expected to come from Mercury. Were licensing policy and regulation to be altered, more entry would be possible. Whether entry will actually occur depends on the profit that an entrant foresees will be generated, and over what period. We therefore think of current licensing and regulation as potentially creating new opportunities to make profits.

The analytical question for the regulator is: Were one or more of the regulatory constraints to be modified, would significant new profitable business be done? The bigger the potential business the more important is the constraint. This should provide the key focus for regulatory analysis. BT can be expected to oppose entry insofar as this will tend, after any competitive move which is open to it, to lower its profits. The policy issue is thus to predict the gain to consumers from entry, and the change in profits to entrants and incumbents which will ensue from a given regulatory change, and to order regulatory changes accordingly. We do not have the evidence to calculate the preferred set of changes precisely. Nevertheless, we can present some firm conclusions.

As Mercury is the current competitor in the principal market, fixed link telephony, future decisions to enter this market must take into account Mercury's progress and prospects *vis-à-vis* the incumbent, BT. The profits earned by Mercury, if any, are the strongest signal to other potential operators about the attractiveness of market entry, together with its rate of progress in penetrating the market both geographically and by total volume. So we first consider Mercury's impact and prospects. We then review major potential challenges to the existing duopoly. Table 2 summarises the results of our analysis.

Table 2:
Potential Competitors in Voice Telephony

Type of Entrant	Prospective Profitability	Competitive Impact
Large private users	High	High
Resale — simple resale	Low	Low
— closed user groups	High	Low
— network managers	Low	Low
PTOs — national	Low	High
— local: cable TV	Low	High
— regional	High	High
Mobile radio	High	High
Satellite	Low	Low

Note: Policy should give priority to entrants scoring highly on both profits and impact.

A. MERCURY'S IMPACT AND PROSPECTS

Mercury's opportunities for profit derive from three principal sources:

1. The high margins available for long-distance and international traffic.
2. The use of lower cost technology.
3. The right of access to BT's network on preferential terms.

The low-cost technologies available to Mercury—microwave and optical fibre transmission and digital exchanges—are in principle also available to BT. Mercury's prospects of establishing itself in the market-place therefore depend on bringing these technologies into service faster than BT can replace its own plant.

Mercury's strategy and priorities have not been made public, but can be inferred from its organisation, services and price structure, and from the regulatory rules that govern its behaviour. The acquisition of Mercury by C&W produced a change in its orientation towards the fulfilment of C&W's long-term global strategy. In terms of services, Mercury has not set out to become the UK's second national carrier. Instead, it has sought to obtain a high volume of originating traffic from a narrow set of customers. These customers are typically:

o highly concentrated in Central London;

o have high long-distance and international calling volumes;

o have relatively sophisticated internal communication systems and management.

The focus on Central London is the result of the concentration there of corporate headquarters and financial services with a strong international calling pattern. Mercury can connect them to its network relatively economically by means of the optical fibre cable, chiefly installed in LHP (London Hydraulic Power) ducts. This is made explicit in Mercury's tariff. Other suitable customers are concentrated in the large conurbations outside London, most of which are located on the main railway routes used by Mercury. Further extension of the Mercury network beyond London and the figure-of-eight serves two distinct functions:

1. to enable direct connection of additional customers or additional locations of existing customers;

2. to carry interconnected calls further (long-distance bypass).

1. Connection Cost Barrier

The primary economic constraint on the extension of Mercury's network has been the cost of local connection of customers. *Microwave* connection to a Mercury node is normally economically viable only if more than 10 circuits are required. As Mercury circuits often carry only outgoing traffic, locations connected by microwave must have sufficient traffic to justify about 20 circuits if the customer is to obtain a significant net saving in the telephone bill (capacity required for other Mercury services can contribute to this total). *Optical fibre cable* can be deployed only if there is a concentration of potential customers or through association with a local broadband cable network. Mercury has now begun to install local optical fibre networks in a few other business centres.

The tariffs charged to customers closely reflect these considerations. Mercury makes a substantial charge for connecting customers to its network. Ordinarily, a customer will already be connected to BT and so, in deciding whether or not to change to Mercury, the additional cost of connection must be balanced against prospective savings on calls. Mercury charges less for connection by cable. For example, the charge for a microwave connection is £3,000 for any number of circuits up to 30. For a cable connection, the minimum charge is £1,264 for 15 circuits, rising to £2,389 for 30 circuits. In addition, rental payments averaging about £100 per circuit per annum must be made. These high fixed costs per location and per circuit contrast with BT's tariff structure, where the rental charge is £90 per annum for the first and each extra business exchange line.

To decide whether it is worth connecting to the Mercury switched voice service, users must also know their calling volume and patterns. The discount offered by Mercury is higher for short calls than for long, for long-distance calls than for local, for some international destinations rather than others, and at peak periods than at off-peak. The calculation of net gain from using Mercury is complex and requires more information than the average telecommunications manager possesses. Depending on the precise circumstances of each case, the net saving from using Mercury might vary between 5 and 25 per cent of the BT bill. In practice, therefore, Mercury can sell its switched voice services to users who have clearly identifiable requirements and destinations for which Mercury is able to offer a substantial saving compared with BT, or who have an exceptionally large volume of traffic originating from a specific site. At the end of 1988, Mercury had about 1,500 sites directly connected to its switched voice service (2100 service).

Through interconnection with BT, customer building can proceed without network extension. But connection via a BT exchange line (required to obtain the 2200 and 2300 services) reduces Mercury's margins on traffic carried. The *2300 service* is designed for single line business and residential customers and so can indicate how Mercury has set out to attract the bulk of users. At current prices, and assuming a normal spread of calls between local, long-distance and international destinations, a residential user would need to have a prospective BT bill of about £340 a year, or £85 a quarter, to make a significant saving by changing to Mercury.[1] This threshold occurs because the user of the 2300 service continues to pay an exchange line rental to BT and, to avoid complex dialling routines, must buy a relatively expensive telephone capable of selecting Mercury at the press of a button. Mercury's 2200 and 2300 services would have a wider appeal to users who make a preponderance of long-distance calls; these tend to be located outside London. By the end of 1988, Mercury had about 5,000 sites connected to its 2200 service and 12,000 customers for the 2300 service.

---

[1] In 1988 Mercury charged £45·20 for the telephone plus £7·50 per annum for an authorisation code per line (exclusive of VAT). It is assumed that the subscriber spends with BT, net of rental charges, 50% of the bill on local calls, 40% on long-distance and 10% on international calls. Mercury's tariff offers on average no saving on local, 15% on long-distance and 10% on international calls. Thus to make a 10% saving on the BT bill over three years, the user must expect to spend £337 per annum with BT.

BOX 2

Mercury's Revenues 1988-89

	£ million
Switched voice services	
2100	47·0
2200	10·0
2300	3·0
Private circuits	40·0
Telex	10·0
Other services	10·0
Total revenue	120·0

Notes: Revenues are stated net of interconnection payments to BT. Cable and Wireless does not publish any information about the revenues of its Mercury subsidiary. These estimates, though probably not accurate in detail, show the great importance of switched voice services in Mercury's business.

2. Mercury's Non-Competitive Strategy

Mercury's strategy is likely to make it profitable within the duopoly period. In its 1988-89 financial year, Mercury achieved revenues of about £120 million (see Box 2). Over 90 per cent of these revenues were generated from sites in the City of London. It could seek a higher market share by offering a larger discount to customers or by extending its network more rapidly. Either strategy would be likely to sacrifice profits in the near future without raising Mercury's market share to a level that would be sufficient to deter all potential entrants by 1990.

For BT, what is most significant is that Mercury has not yet posed a serious challenge to BT's profitability, in particular to profits earned from BT's major customers. Mercury appears content to offer these customers a small discount on the telephone bill, namely, lower call charges which may be partly offset by higher access costs, and so to build up the number of customers slowly. BT has not repeated the tactics of targeting potential Mercury markets which, as we saw, the original decision to license Mercury prompted. In 1986 and 1987 it considered countering Mercury with discounting to large customers by the '*Optional Calling Plan*' (OCP—see Glossary). This weapon of retaliation was, at least temporarily, sheathed. In any case, had it been implemented, BT would have faced a strong risk of net loss of profit because, to avoid running afoul of Condition 17 of its licence, all large customers would have had to have been offered the OCP, not merely the relatively few at any one time likely to become Mercury customers.

The market for voice telephony is likely to grow faster under the duopoly than it would have done had a policy of limited liberalisation not been pursued. But Mercury's market share will grow at a rate that does not entail great sacrifice by BT. BT has not had to make exceptional efforts to improve its costs. With consumption in the UK still growing at 7 per cent per annum in volume terms, extra capacity for fixed link telephony beyond that likely to be provided by the duopolists could be remunerated. If so, openings for more network operators may be created. We now examine the possibilities.

B. LARGE PRIVATE USERS

Duopoly and other restrictive legal barriers to entry encourage customers to substitute their own private supply. Large firms that are intensive users are particularly likely to respond in this way. Own account operations are often permitted provided they do not spill over into public provision. For example, the UK regulated road freight transport in the early 1930s, and deregulated it in 1969. By that time, own account operations, either through direct ownership or by contracting to others, had become substantially larger than public operation for hire, using whatever scope the licensing law allowed. After deregulation, own account operations were reviewed. Some became public carriers. Over the years since then, most companies have hived off freight transport to the now fully competitive public carriage operations. So the sequence in transport was: regulation; evasion of constraints by own account operators; deregulation and redivision of the market between own account and independent supply.

Telecommunications in the UK is at a stage when own account operation could emerge as a serious competitive threat to PTOs. Of course, telecommunications has some distinctive characteristics, for example, economies of scale which are almost wholly absent from road freight transport. Yet there can be little doubt that by far the most important dynamic in altering supply conditions is where large, multi-location firms that are intensive consumers of telecommunication services place their business.

What are the prospects for competition arising from large private users of telecommunication services? We consider

(a) how exogenous influences are likely to affect the inducements for large firms to create their own telecom services;
(b) how liberalisation has affected these inducements so far; and thus,
(c) the basis for a prediction of which regulatory changes will most affect those inducements.

Private networks may consist of facilities leased from *PTOs* or privately provided. The facilities may be owner operated or under professional management. At this point we do not distinguish between private networks including leased lines and those that are privately constructed and run. In the concluding part of this section, we take up the question in the context of the regulatory changes required to open up market entry to private networks.

1. Own Account Operation

Competition can arise from the extension of private networks in two ways. First, a major user can substitute own account operation for telecommunication services bought from PTOs. This may well involve development of its network either by further leasing capacity or outright purchase of facilities. The other way in which competition could arise is by private networks carrying traffic for other users. In a sense, therefore, they would then cease to be private, becoming direct competitors of PTOs. This step would normally be associated with network development and other changes in activities undertaken by the firm.

Own account operation represents a more certain demand than does carriage for others. For this reason, internal demands for telecommunications are likely, at least initially, to be the leading factor in the decision. In the USA, private provision of transmission capacity was the first telephone activity to be liberalised (the 'Above 890' decision of the FCC in 1959). A long-term effect of this decision was to enable firms with sufficient 'own account' demand to construct efficient networks using either privately provided or leased circuits. The scope for exercising such a choice is still denied to UK firms. 'Own account' developments are therefore much more advanced in the USA than in the UK. In the USA, many long-distance microwave networks have been in operation for some time. Most consist of leased transmission facilities and are managed by AT&T. There are some 300 private networks under AT&T management.

US Developments in Own Account Operations

In the USA, own account operations have been the leading source of competitive developments. The 1987 Huber report,[1] very important in the US debates on telecommunication policy, did not recognise this, but its evidence strongly points to it. Huber shows, firstly, that the number of PABX connections has grown much faster than have direct exchange lines. Between 1982 and 1986, the number of residential exchange lines increased by 8·8 per cent and business single exchange lines by 33 per cent; the number of PABX trunks and terminals doubled.[2] Secondly, the number of *Centrex lines* has declined, although the figures are disputed. Centrex may be regarded as a bid by the public networks to contain privately operated PABXs. Shared tenant services, which from the perspective of the user are collectively financed independent supply, have not been successfully established during this period (market share less than 1 per cent in 1986). Huber attributes this lack of success to regulatory restrictions maintained at the insistence of local public network operators.[3]

In terms of the sales of telephone equipment, revenue from PABX sales have overtaken those of new central office equipment. More generally, private buyers now account for 80 per cent of sales of satellite transmission services, 40 per cent of telephone switching, 20 per cent of microwave transmission equipment, and 20 per cent of optical fibre cable and associated electronics (Huber Report, p. 1.11).

In areas served by Regional Bell Operating Companies (RBOCs), private microwave transmission links increased from 0·3 million in 1982 to 3·4 million in 1986.[4] The notional capacity on private fibre systems has reached 250 million circuits. Presumably, only a fraction of this capacity is in use. These optical fibre systems, such as the new Metropolitan Area Networks or *MANs*, are termed private but are not solely concerned with own account operations as we have defined them. Nonetheless, optical fibre systems provide a stimulus to own account operations by 'unbundling' network services. MANs are particularly significant in providing access from dense urban areas to satellite earth stations located on the edge of cities (teleports), which are predominantly own account. The installed capacity of earth stations has doubled since 1982.[5]

The broad picture is clear. US firms that are large users of telephone services have moved quickly to take advantage of the

---

[1] *The Geodesic Network: 1987 Report on Competition in the Telephone Industry*, US Dept of Justice: Antitrust Division, Prepared by Peter W. Huber.

[2] Huber Report, Table L.4.

[3] Huber Report, Table L.10.

[4] Huber Report, Table L.15.

[5] Huber Report, Table L.18.

new opportunities. Private networks have been reconfigured to increase own account switching and transmission. The trend towards own account in switching is clearer and probably more significant. But interexchange carriers, particularly AT&T, have sought to encourage direct connection of large users. Connection could be by leased circuit or private transmission (microwave or optical fibre). AT&T, MCI and US Sprint offer resale (local by-pass) and Centrex-type services, installing multiplexers (see Glossary) on users' premises for the purpose. In effect, major users now have a duopolist to compete with each RBOC.

In the USA the large user has, in recent years, gained in many ways: an ability to buy long-distance connection more cheaply, with no offsetting constraints on its rights to use public networks for local collection and delivery of calls. The price of using the latter facilities is still strongly influenced by regulatory arrangements, encouraging large customers to keep to a mixed strategy of 'own account' and use of public networks. In particular, the rebalancing of PSTN charges has not been permitted to proceed to the extent of increasing local call charges significantly.

2. Motives for Own Account Operation in the UK

The potential value to firms of using telecommunications services to improve internal efficiency is almost certainly rising both now and in the longer run. For example, sophisticated data manipulation and communication among customers and suppliers are growing rapidly. Distinctions between voice and data messages must, in the interests of efficiency, become less rigid (this tendency is linked with, but separate from, the technical capability to integrate voice and data in a single digital bit stream).

The proliferation of technological options increases the firms' potential to adapt to changes in the market-place. For large users contemplating expanding their own supply, the most important consideration is that choice is widening and costs of equipment are falling. Fibre optic cable and satellite services offer ever-widening options, and microwave has by no means exhausted its competitive attraction. Private networks, as opposed to the very much larger public ones, may be able to tailor demands on technology with more discrimination.

The question is whether these developments broadly favour own account operations in the UK. To answer this question, it is necessary to determine how costs will affect all telecom providers. The economic balance between public and private networks depends on the continuing significance of economies of scale and scope (see

Glossary) and on shifts in the relative costs of transmission and *switching* technologies. Both the initial costs and the operating costs of networks are falling in real terms, with the result that economies of scale and scope are becoming less important. The cost of transmission is falling faster than the cost of switching. Direct comparison of the costs of public switched services, leased circuits and private systems using microwave transmission indicates that there is a wide range in which each is the most economic.

The evidence presented in the CSP International report, *Deregulation of the radio spectrum in the UK*,[1] indicated that private microwave radio fixed links would cost substantially less than leased lines for links over 10 kilometres and comprising more than 10 circuits.[2] Short-distance links by cable have also been shown to be capable of being more cheaply provided privately than by the PTOs, which of necessity route cable circuits through central nodal points. This evidence suggests that the most economic arrangement for private networks would be a mix of leased and privately constructed links. At present, this is not allowed.

In the provision of switching (i.e. various forms of exchanges), economies of scale are not strong. Public exchanges and PABXs, which provide switching for PTOs and private users respectively, occupy different market segments, but are very similar in capabilities and cost. In recent years, suppliers of PABXs have adopted and marketed digital technology at a more rapid pace, apparently because of the greater complexity of software development for public exchanges. The introduction of 'Centrex' as a method of providing switching within public exchanges for private networks now offers users a direct comparison with operating their own PABX. The reconfiguration of large firms' own networks, together with developments in switching, offers numerous, specific opportunities to lower costs. Private networks may also simply be more efficient. For example, the benefits from conversion to digital technology may be secured more speedily on a smaller scale. Moreover, private networks will probably be able to drive better labour bargains than PTOs can. All this favours own account operation, and can be expected to continue.

Since 1984, however, the cost of buying in services has also fallen. Tariff rebalancing between switched and leased services and between long-distance and local services has been adverse for constructing private networks and has probably left many over-

---

[1] CSP International, *Deregulation of the Radio Spectrum in the UK*, London: HMSO, 1987.

[2] *Ibid.*, Exhibit 4.3, p. 41.

extended. Our analysis of the duopoly indicated that since Mercury was able to establish an average discount of about 17 per cent from BT's call charges, the cost to users of external supply of telephone service has been reduced. But over the longer run, this once-for-all shift in favour of PTOs will be exhausted.

Duopoly Policy Reduces Incentives for Own Account Operation

The effect of the duopoly policy has been to lessen the attractions of own account operation. This assertion implies a judgement about how free the choice was in 1980 under the old régime. Leased line charges were then kept low. Regulation was operated entirely by BT; some concessions to large firms had been made. Indeed, it was probably true that a well-organised private customer could get useful dispensations from BT, though falling short of expanding own account operations to take on outside switched voice business in competition with BT. The number of such dispensations was also limited by the zeal with which radio frequencies were conserved. Specific examples of private voice telephony networks linking more than one legal entity included the Stock Exchange and firms with subsidiaries linked to their network.

Since the 1984 Act came into force, own account operations have been governed by the Branch Systems General Licence (BSGL). The policy underlying this Licence was to codify the rules previously applied by BT, without significantly liberalising them. This policy was to prevent the connection of privately provided circuits to public networks and to limit the carriage of calls over leased circuits where public networks provided an alternative. Relaxation of the rules for leased circuits would follow when BT had increased its charges for leased lines. No commitment was made about privately provided circuits.

As BT's custom had been to grant *ad hoc* concessions, codification left a large number of firms in the anomalous position of having a more liberal inheritance from BT than was now on offer from the Government. The problem of these 'grandfather' licences was resolved by issuing over a hundred temporary licences which preserved for the time being what BT had permitted. In 1987, when the BSGL was revised, the first, tentative liberalising steps on the use of leased lines for voice telephony were taken. The change made was to permit more efficient designs for private networks composed of leased lines, while retaining the restrictions on potential competitive developments. However, as BT had in the years up to 1987 increased charges for leased lines substantially, the net effect of regulatory change was to reduce the demand for leased lines.

These developments provide the background to the proposed relaxation of the use of leased lines contained in the Consultative Document issued by Professor Carsberg in February 1989.[1] The Consultative Document acknowledges that current restrictions appear arbitrary and present continuing difficulties for users and for equipment suppliers in understanding what can and what cannot be done. The main changes proposed to the BSGL would permit unrestricted routing of calls over leased circuits from one site to another, such as call diversion, while retaining restrictions on wholesaling activities (resale). A separate licence may be issued for resellers.

In terms of the feasibility and cost of developing one's own system, it might seem that the BSGL's intended liberalisation of switched voice within a group would be a fillip. This has been dampened by the obligations under the network code of practice that such a choice would entail. In practice, the standards imposed on a private network are onerous; the degrees of freedom permitted in CCITT standards are almost entirely taken up by the PTOs or, more precisely, by BT. This not only imposes a high cost of compliance, it also forecloses one important dimension favouring own account operation—the possibility of adopting much lower and therefore cheaper service standards internally.

With tariff rebalancing, an important potential effect on phone bills is the possibility of profitable local bypass of BT in particular, especially on voice. The 50-metre rule for extension of own account operation was, if anything, a regression from the *de facto* situation before 1980. Then, it was possible, subject to BT's consent, to extend private wiring within the local call charge area. What constrained development was not *de facto* regulation so much as the costs and relative advantages of bypass, especially in London.

The new 200-metre rule does not restore the situation; with price changes, the pay-off to constructing one's own network is greater but, in practice, more constrained than in 1979. Moreover, since the ban applies equally to all alternatives to cable (terrestrial and satellite microwave) the flexibility of being able to choose a mixed strategy in building private systems is foreclosed.

Equally, the prohibition on connecting privately provided and leased circuits has been restrictive. Since sensible own account networks would nearly always incorporate some leased lines to greater or lesser degree, regulatory clarity about the inhibitions on switching voice over leased lines has foreclosed a partial strategy incorporating them.

---

[1] Director General of Telecommunications, *Further Deregulation for Business Users of Public Telecommunications Systems*, London: Oftel, February 1989.

The adverse regulatory conditions now faced by own-account operations could, of course, be removed with appropriate reforms. If users were permitted to mix leased and privately provided circuits in their PABX networks, but not resale or shared use, there is little doubt that substantial expansion in large users' networks would occur. If, in addition, it was made clear at the same time that there was a firm future prospect for unlimited development of outside business, own account operation would be further encouraged. On the other hand, the immediate liberalisation of resale and shared use alone would produce a different reaction. Firms would foresee that they might eventually be running more of their own capacity, but would extend their networks with caution. They would wait and see what specialist networks emerged after liberalisation.

C. RESALE

Resale is the generic term for a range of telecommunication services, the common element of which is the wholesaling of a basic service offered by a PTO. Users value wholesaling activities which offer them a lower price per unit for the same service as provided by a PTO, or which tailor PTO services more closely to their requirements in terms of delivery times, quality or maintenance and other support services ('network management') or which involve adding value to their basic service. Wholesaling is virtually universal in industries outside telecommunications, including service industries. There is no doubt that the phenomenon would arise in telecommunications with free entry. Under the duopoly policy, wholesaling except in relation to value-added services is not permitted.

1. Simple Resale

The archetypal wholesale service is termed 'simple resale'. It involves the business of gathering traffic from one local network, carrying it over a high capacity leased line and delivering it to another local network. Simple resale would be very similar to Mercury's 2200 service for indirectly connected customers. All licences issued under the 1984 Act contain a blanket prohibition on this form of competition. The prohibition may be lifted in 1989 as part of a general relaxation of controls on PABX networks.

Two elements—attracting customers by lower prices and reducing the cost of supply—basically determine the margins available to attract entry. The first analysis of the margins to be obtained from simple resale was in the Beesley Report. There it was demonstrated that, for a five-minute peak-rate call, a reseller could expect a gross margin of 20p on short national routes and £1 on long national routes.[1] The analysis was undertaken at 1980 prices. At 1988 prices, the equivalent margins are:

    Long national (b)   18p
    Long national (b1)   9p
    Short national (a)   6p

Although substantially reduced by tariff rebalancing and inflation since 1980, the gross margins may still be feasible for some long national routes.

The reseller's costs of leasing transmission capacity have also changed as a result of BT's tariff rebalancing. Not being a commercial possibility, we presume that it has not been the focus of BT's rebalancing calculations. Rather, to the extent that competitive concerns have motivated the changes, these have been part of the response to the services delivered over Mercury's network and the optical fibre and microwave extensions to users' premises. That is, the prices offered for analogue leased lines (in which there is no competition from Mercury) have risen relative to digital leased lines. Similarly, the rate of increase of leased line charges with distance has been reduced. The reseller may also be able now to lease circuits from Mercury.

BT's Monopoly Prices and the Reseller

But the reseller, in order to build a substantial business, must lease a quantity of capacity for which prices are, in effect, not subject to regulation. This raises the prospect that BT might seek a premium on high-capacity lines connected to its PSTN. The long-running regulatory debate over the pricing of BT's access lines suggests that the issue will not be easily resolved. The access line discussion has run in terms of:

(a) treating access by PTO customers as a separate matter from interconnection of networks;

(b) seeking a cost-justified price for access lines.

For users of PTO facilities who embark on resale there is no business distinction between access and interconnection. The critical question is the level of charges. It might be accepted that

---

[1] The basic calculation compared the price of one long-distance call against two local calls.

resellers should not get terms from BT which equal or better Mercury's interconnection terms, if they escape PTO obligations. But a decision on the 'just' price will be elusive. If a feasible commercial solution is found, as has proved to be the case with Mercury, it will still be arbitrary. By contrast, an approach would be to permit BT to charge what the market will bear for high-capacity links but to insist that there be no discrimination between resellers and others.

Helping to constrain BT's pricing for access in a manner adverse to resale is that, on many routes, capacity could be leased from Mercury. Mercury's tariff for leased lines is not structured in a way that permits direct comparison with the figures given in the Beesley report. As already noted, the high fixed elements in the tariff discourage low volume users. If a reseller chooses a high-density route and can gather sufficient traffic to fill 30 circuits (the capacity of a 2 Megabit circuit), then, at current prices, the annual cost per circuit leased from Mercury for the 112 km. route used as an example in the Beesley Report could be as low as £644, compared with £1,520 at prices current in 1980.

A complete analysis of the prospective profits from simple resale would need to take into account the costs of switching at either end of the route, additional marketing and billing expenses and the size of the discounts relative to BT's and Mercury's prices for long calls. There may also be significant costs of compliance with regulations to be taken into account. Nonetheless, it has been shown that simple resale has not been eliminated as a source of competition.

The market for simple resale could develop as an unconnected series of route-specific operations tapping particular concentrations of business or as competition among a few firms each offering long-distance bypass in a number of geographic areas. Much would depend on the nature of the licences awarded to resellers. Resellers would not be PTOs and so should avoid PTO obligations. The Consultative Document issued by Oftel in February 1989 hints that they might be surcharged in order to contribute more to BT's profits.

Among the customers which might be attracted by a resale operator are users with a high frequency of calling within a group or to a particular destination. These would include many large customers of BT and Mercury who, as we have argued, may well also consider, under a liberalised régime, substantially increasing own account operations. From the reseller's point of view, use of the PTO switched voice services at either or both ends of a call could then be eliminated, saving the local call charges. Margins for resale could then be more attractive. Building a set of customers of this kind would also bring economies of marketing and billing expenses. A natural way to build a resale business would be to put together such favourable customers and then add smaller, lower-margin customers. From the regulatory point of view, market entry via a closed user group can be considered as a first step towards liberalising resale in the voice market.

2. Closed User Groups

A 'closed user group' is a set of customers for whom services are provided in common and exclusively. Some closed user groups, of the following kinds, are already permitted:

o   under the BSGL's complex conditions;
o   if already in existence in August 1984, for example, the Stock Exchange's network;
o   if the members have a common business interest in communication and their purpose and effect is not to divert revenue from the duopolists.

These limited permissions do not permit an extension of competition. In particular, the strict limits currently in force on the number and use of bilateral private circuits prevents the building of significant closed user groups. Closed user groups formed for data communications could economically add voice traffic to their network. At present, the restrictions on charging for calls carried over data networks prevent the effective exploitation of such economies of scope.

As PSTN interconnection would substantially increase the traffic potential of a resale network, we expect closed user groups formed for voice services to survive after liberalisation only in circumstances where simple resale is infeasible. A particular possibility would be for local networks in such areas to gather traffic to be delivered into London by shared use of out-of-area exchange lines. Such a network was proposed in the early 1980s for Cornwall, primarily as a means of encouraging the location of firms on sites owned by English China Clays. The specific attraction would be for such firms to have the appearance of a London area number. The introduction by BT of 0800 and 0345 services may have expanded the market for such services.

3. Network Management

Restrictions on the use of leased lines in the 1984 Act licences also prevented the emergence of an independent sector of professional network managers for basic voice and data services. A few firms have entered the market, such as EDS which manages the networks of General Motors and Unilever. However, within the UK, it is not possible for network managers to offer significant economies by gathering traffic from several companies. Under present regulations, management of private networks in the UK remains as it always has been, primarily an activity for which the main qualification is the ability to get on well with BT staff, in order to ensure a reasonable standard of service. There is little scope for professional management to offer savings in the telecommunications bill. Where company networks extend overseas and so are engaged in discussions with a number of foreign operators, there may be gains from employing specialists to conduct negotiations.

International data transmissions require particular management skills and several companies have established profitable businesses in this sector. Having secured permissions and connected links, it is sensible to manage voice traffic as well, even though under present regulations, the margins are small. Further economies could be offered to users if their traffic could be aggregated; this appears to be contrary to CCITT (see Glossary) recommendations. For the UK, the business at stake here is the location of data processing activities and the communications facilities of international companies. The UK currently has a strong position; to sustain it as foreign network operators improve their offerings requires progressive liberalisation.

If resale is permitted in various forms, the current choice for private network operators—to lease or to use the PSTN, BT or Mercury—will become more complicated and, equally important, the terms of choice will be shifting more rapidly. Specialists in measuring traffic and calculating the trade-offs can offer to make this choice on a continuing basis by managing private networks for a fee. By comparison with consultancy arrangements, typically used at present to configure private networks, professional network management involves both some risk sharing and the possibility of sharing the benefits of economies of scale. Network managers would themselves invest in switching and network control equipment and lease transmission capacity to be shared among their customers. Firms unwilling to invest in an own account operation may find the offer attractive. In these circumstances, our judgement is that network management involving the resale and shared use of leased lines could prosper in the UK even without extensive international liberalisation.

4. Implications

Resale is still a competitive option at current PSTN call charges and leasing costs. We think the relevant PSTN prices are stable and so the option is unlikely to disappear as soon as it is permitted. Leasing costs are not stable and will require regulatory attention. Resellers require the protection of a non-discrimination rule on leased line prices charged by BT and Mercury.

D. A THIRD PTO?

A national network with public service obligations is the only form of entry that is wholly consistent with present policy. It is also the least likely. On the basis of Mercury's experience, a potential entrant might well conclude:

o   The key to positioning for entry into fixed-link telephony is access to sites for microwave transmitters and to rights of way for cables.

o   Mercury may eventually be a commercial success, but the time taken by Mercury to reach profitability and the continuation of manifold regulatory disputes indicates considerable uncertainty. Increasing the number of participants in an oligopoly is also destabilising.

o   Mercury appears to be an efficient, albeit slow moving, supplier, drawing its resources from Cable and Wireless. Other entrants, if precluded from using foreign expertise, would probably face higher set-up costs.

o   Mercury's customers appear to require a substantial discount (more than 10 per cent sustained over several years) to contemplate changing from existing sources.

Three Potential Contenders

Given these considerations, only three possible entrants are well placed to enter by virtue of their ownership of national networks used for other purposes—British Gas, British Rail and the CEGB. Of these, British Gas is the front runner for two main reasons:

1. It already possesses a microwave network, albeit of limited capacity (2 Megabits), extending beyond Mercury's reach, and has experience in running modern communication systems. Much of its entry costs are already sunk.
2. It is privatised, which we must reckon a prerequisite for a public

utility candidate. Being very profitable and subject to regulation on its principal activities, it can view with equanimity a strategy of rapid, large-scale investment with a postponed but suitably impressive pay-off.

British Gas, alone or in partnership, could enter tomorrow, if permitted. Its preferred strategy might be to add to its existing capacity gradually as its customer set built up. Its headstart in owning a national network gives it the advantage of a flexible investment strategy.

By contrast, British Rail is not on the list for privatisation in this Parliament. To allow it to enter on its own as a nationalised industry would reawaken the dormant debates over private/public sector boundary conditions and over legitimate sources of capital for nationalised industries. These were the puzzles that earlier helped to induce BT's privatisation. In BR's case, we see no real prospect of their being resolved. But it is well positioned to enter with respect to its owning rights of way, geographical spread, engineering expertise, and considerable access to potential customers. BR might be a partner in a joint venture.

The CEGB is in process of privatisation, is profitable and has rights of way. However, its management has had other preoccupations, particularly recently, and lacks experience both in running communications networks and in dealing with final consumers. In several countries, telecommunication services and electricity have been supplied by the same company as part of distribution. The intended post-privatisation structure of the UK electricity supply industry has produced separate area and central distributors. This is not the most favourable outcome for competition in telecommunications as a national network. To do this, a jointly owned telecommunications service company for the industry as a whole would be required. If such a development were to take place, it would be after 1990. Meanwhile, the individual privatised area organisation might well be more interested in more local forms of entry, considered below (pp. 64-67).

British Gas Linked to Mercury?
If British Gas decided to enter, it could form a mutually profitable arrangement with Mercury. The two networks would have strongly complementary market functions, British Gas extending Mercury's domestic potential in return for delivering extra traffic to Mercury's international routes. British Gas is well placed to avoid the problems over making a deal which cable TV franchisees now encounter (see below, p. 63). The precise form of the association would also

doubtless be greatly influenced by regulatory conditions at the time. Specifically, Mercury would have to consider what it might gain once control of the national numbering plan has been wrested from BT.

For these reasons, British Gas can be expected, if not to seek, then certainly to welcome a Long Line PTO licence offered on equivalent terms to that granted to Mercury. Indeed, British Gas could contemplate a licence identical to BT's. We do not think there could be any other takers for such licences before 1991. That is to say, because British Gas is the only plausible candidate and has only very recently moved into a position from which entry would be feasible, the duopoly policy has so far not, in reality, actively excluded other potential national networks. As we shall see in the next section, it may have excluded regional and local networks.

It is unlikely that, by 1990, the position will have changed. With Mercury moderately successful and British Gas uniquely placed to enter, if not already doing so, other British candidates should be deterred.

The position is materially different for certain foreign-based network operators, namely, those that command a large home market and international access. For these, reciprocal entry agreements could be attractive. A major European network could offer expertise and deliver a very substantial overseas potential to complement a British presence in their market. US networks are in much the same position. The bargaining, at industry and governmental levels, required to extract sufficient reciprocal concessions in return for ownership rights for a PTO operation might be too daunting for foreign operators to contemplate. In any event, foreign participation in a British PTO may well remain politically unacceptable.

E. LOCAL NETWORKS

When the duopoly policy was announced, there appeared to be little scope for competitive market entry into local network operation, at least on PTO terms. Connecting users to exchanges is highly capital intensive, even after making allowance for BT's high costs. In the USA, the model of an efficient telephone industry when the duopoly approach was decided upon, AT&T's willingness to be divested of its local networks appeared to confirm that profits lay in long-distance and international services. At BT's prevailing prices and costs, and not withstanding the example of Hull, local distribution that included an obligation to connect residences as well as businesses was not profitable.

Mercury was, of course, free to enter local distribution, but expected to confine its interest in direct connection to major customers, using interconnection with BT's telephone system to broaden its customer set and to deliver most calls. In the event, Mercury has found that the operation of local networks can be economic in city centres. It has few, large digital exchanges and, especially, connects customers with high traffic levels by optical fibre. Customers who cannot be served profitably are deterred by the structure of Mercury's tariff. At the same time, new techniques of local distribution are being developed that are appropriate to customers with lower levels of demand for service. Nonetheless, there remains little prospect of Mercury constructing local networks throughout its licensed area.

Since 1983, BT's charges for rental of exchange lines and for local calls have risen both relatively and in real terms. Charges for other services provided by local areas have also risen substantially. If, as appears to be the case, BT's local networks are making a profit on fully allocated costs, then market entry has become feasible. If, as also appears to be the case, other operators starting afresh could have substantially lower costs and the possibility of selecting their customers, as Mercury has done, entry into local distribution may be attractive, particularly to companies in possession of rights of way or of cable already laid in the ground.

The broadband cable networks run by cable television companies have been identified as the main candidates and we analyse their prospects in detail. They are not the only potential entrants into local distribution, particularly in urban areas. Other duct networks exist in most cities, laid many years ago for tramways. London Regional Transport is particularly well placed, having the Underground as well as tramway ducts. The London Hydraulic Power Company had an extensive network running through the City and the West End, but has already been acquired by Mercury. Private networks laid in ducts exist, although on a more limited scale. None of these have obtained any permissions. The value of rights of way depends on telephone demand, how far they are substitutable at particular locations, and the costs of allowing fresh uses to be made of them. Little deters owners of rights of way in the public sector such as London Regional Transport from offering their capacity, if demand is there. Mobile radio technology is another source, considered below (pp. 67-69).

1. Broadband Cable

In the early 1980s, broadband cable networks were encouraged, as a way of distributing additional television channels and of developing new forms of data communications (termed 'interactive services'). Broadband cable was also seen as a potential source of competition in telecommunications. The profitability of commercial television was expected to be the inducement for the private sector to invest in these networks. Interactive services and voice telephony could then be provided relatively cheaply.

Nevertheless, even before the duopoly policy was in place, the Government determined that only BT and Mercury would be allowed to offer voice telephony. BT and Mercury were given exclusive rights to provide data communications in business centres as well. The reason given for this decision was

> 'the prospect of BT's and Mercury's financial position being eroded in the event of interactive services depriving them of local and national traffic'.[1]

It is difficult to understand how so great a miscalculation could have been made. It was never envisaged that broadband cable networks would carry long-distance traffic. In any event, a separate regulation gave BT and Mercury the exclusive right to all transmissions between cable networks, television as well as telecommunications. So national traffic was not at risk. And local traffic was not profitable. Why would broadband networks, primarily concerned with television and known to have high costs and construction times, even seek to compete with BT?

There is further evidence that the processes of market entry were not understood. Broadband cable networks were given limited franchise areas—typically 100,000 homes (well below the average size of BT's local call charge areas). They were also given service obligations of a kind considered appropriate to monopolists—a timetable to build their networks and a universal service obligation. The standard was five years to pass 90 per cent of the homes in their franchise area, with intermediate coverage targets as well. None of those licensed has been able to meet any of the targets set. None has yet made profits.

The risk that BT might come to dominate cable television was perhaps more realistic. Much of the cost of building broadband cable networks lies in the ducts and customer connections, where BT has assets already in place. The prospect of duct sharing may explain why BT was not expressly prohibited from the cable television business. However, the Cable Authority has indicated that BT would not be allowed to monopolise cable television: its current market share amounts to no more than about 30 per cent. Where

---

[1] *The Development of Cable Systems and Services*, Cmnd. 8866, London: HMSO, 1983, para. 185.

BT is involved, the broadband cable network does not offer any telecommunication services.

Mercury is interested in using broadband cable networks to extend its services beyond the central business districts where it has established a presence. With so few cable networks in the field, however, Mercury cannot rely on them to add significantly to its business.

Tough Task for Cable Operators

With the approach of the review of the duopoly policy, other investors might be attracted, in the hope that the outcome of the review would be favourable. The Director General has repeatedly encouraged hopes that, after 1990, broadband cable will be taken more seriously as potential providers of local telephone services. The recent interest of US telephone companies in investing in broadband cable in this country may be based on such hopes.

Even with a revival of investor interest in such networks, there are formidable problems ahead. Only coincidentally will a network laid out to serve the residential television market also be appropriate for business telecommunications. It can be safely predicted that getting necessary permissions will be difficult. Mercury, even with the experience of Cable and Wireless and the privileges of a duopolist, is apparently permanently ensnared in complex and obscure regulatory battles to establish a profitable business. Broadband cable network operators are a fragmented group with no experience in telecommunications in this country. They will have an even tougher task.

A key issue is the division of margins on interconnected calls between the local network operator and the Long Line PTOs. The arrangements between BT and Mercury, as determined by Professor Carsberg, left little margin for local calls. That is why Mercury does not carry local calls for indirectly connected customers. Tariff rebalancing, particularly as it leads to BT charging higher rentals, is gradually improving the margins available. Calculations that we have done indicate that the basis for a profitable deal with Mercury do now exist.

If present regulations persist, however, it is unlikely that the use of cable networks to provide voice telephony services will develop. The market areas over which local telephone connections and calls will yield margins are much wider than cable television franchises are ever likely to be. Broadband cable networks must be allowed to deal direct with each other and to link with own account operations and with resale networks once they are liberalised. By such means it may be possible to raise prospective profits to the level necessary to induce investors to undertake the significant risk involved in entering the voice telephony market.

The Broadcasting White Paper contains a very different analysis of the prospects for local cable networks.[1] What it says is rendered unbalanced by the tactical silence on the telecommunication possibilities (to have said anything of substance would have put at risk the Government's intention not to review the duopoly policy until November 1990). The White Paper remains concerned with the problem that local distribution networks are potential monopolies, despite the evidence that prospective profits are not sufficient to promote TV-based entry. It therefore proposes a package of substantial relaxations in franchising rules, combined with a new, draconian restriction on what franchisees may do. The new incentives are:

o removal of onerous technical standards suitable for interactive services;

o larger franchise areas;

o less demanding coverage obligations.

To balance this, franchisees are to be limited to the delivery of services prepared, packaged and retailed by others. This proposal reveals again a lack of understanding of how market entry can proceed. Broadband cable networks must build up their business by selling access to services. Ordinarily, these services will be provided by others. For cable television, there is already a clear division between programme providers and network operators. A separation does not need to be enforced by regulation. For voice telephony, broadband cable networks could be retailers of Mercury's long-distance and international services. Alternatively, Mercury could market the services and contract with cable networks to deliver them. Why should the regulator dictate the terms of this commercial relationship? For local telephone service, the distinction between retailing and delivery simply makes no sense.

2. Regional Networks

We doubt whether it is sensible to expect entry into running local telephone networks to conform to the existing structure of BT's local operations. If Mercury's limited ambitions continue to be

---

[1] Secretary of State for the Home Department, *Broadcasting in the '90s: Competition, Choice and Quality*, Cm. 518, London: HMSO, November 1988, paras. 6.32-6.41.

acceptable to the regulatory authorities, then other entrants should not have onerous PTO obligations imposed upon them. As noted, there is then also the basis for complementary developments between Mercury and new local networks. These would contribute long-distance and international traffic to Mercury; in return, Mercury could assist with ancillary services such as 999, directory enquiries and even billing. In the absence of strict regulatory controls on them, we would expect new local telephone networks to have a regional character.

By a regional network we mean a *Long Line PTO* operating in a limited geographical area of the UK. This would comprise a regulatory category which has not yet been used, though consistent with the rest of the licensing structure. Regional operators could offer public services in areas not directly served by Mercury. The areas that are relevant include, in England, the South Coast, South West, North and East and the whole of Scotland, Wales and Northern Ireland. Apart from Northern Ireland, all these areas have a limited Mercury presence, but not an effective alternative to BT.

Opportunities to enter would depend on the accumulation of rights of way or transmitter sites. The public utilities could cooperate with access to rights of way. As we have noted, British Gas would be in a particularly favourable position to contemplate making use of its own wayleaves. In many parts of the country, transmitter sites have been identified and brought into use for mobile radio.

In switched voice telephony, the basis for profitable market entry at the regional level is the fact that BT's tariff for long-distance calls does not distinguish between calls between local call charge areas the centres of which are more than 56 km. apart. A call from London to Reading is priced the same as a call from London to Edinburgh. As the costs of building a network do continue to increase with distance, albeit less than in the past, margins may be higher for calls between population centres and their hinterland than for calls between centres distant from each other. This element of BT's tariff has survived competition from Mercury; while some adjustments to the boundary between 'a' and 'b' routes could be expected to occur in response to entry by regional networks, BT's room for manoeuvre would be limited by its network configuration.

Following the example of Mercury in planning for a digital network rather than the traditional public network configuration, a whole region could be served by one or two strategically located exchanges. In this way, the investment required to enter the market would be substantially less. It would be hoped that the other major element of Mercury's investment, the cost of clarifying regulation, would also be minimised, if only because Mercury would have paved the way.

Tariff rebalancing has made local bypass of BT a potentially profitable option for Mercury. However, the scale of investment needed to cover the country is daunting, despite the prospect of lower-cost technologies being introduced for local distribution. Competition in local networks could be assisted from several sources, as we have just argued. Regional PTOs could be expected to make deals with such networks in order to broaden their customer set.

A Third PTO Option?

Such regional networks are, to some extent, a new conception of what a PTO might be. Because Mercury's licence gives permission to operate throughout the UK, such operators could collectively amount to a 'third PTO' option. A regional network would include major towns where local connections could be installed profitably. The network could also obtain revenues from trunk calls—short national but also in most cases long national calls. The nature of its PTO obligations would also be modelled on Mercury's rather than BT's. That is, it would be able to concentrate on building its customer base among businesses and other high-volume users.

The interconnection arrangements would almost certainly follow those between BT and Mercury. Mercury, or the third national network, would be willing to offer interconnection but would not be likely to accept worse terms than were on offer from BT. As in the case of the national network option, there would be incentives on both sides for them to reach a general commercial agreement, especially with Mercury. The operators would reduce their set-up costs and obtain immediate marketing advantages; Mercury would secure through interconnection a higher share of total originating traffic. Mercury would also retain the right of direct access to those customers in each region most suited to its service offerings.

So far as we are aware, there has been little work done on estimating the prospective profitability of entry into local networks. With the assistance of Dr Peter Gist of the London Business School, we have recently simulated possible entry into part of North-East London. We assumed the network (NEL) would offer a voice-telephony-only service, independent of broadband cable. Our analysis indicates that at the margins likely to be available, with connection via Mercury to provide a full range of voice services, entry is certainly feasible. The simulations indicate relatively low initial outlay, a cost structure which does not throw much weight on lumpy investments and hence rather small vulnerability to differences in the ultimate penetration rate achieved. The principal risk appears to lie in unforeseen delays in connecting customers, rather than in the number of customers eventually achieved.

This work is only a start to the analyses which need to be performed to judge the possible scale of entry. Important extensions would include estimates for less favourable areas than North-East London, which contains a high density of potential customers; would consider entry at different scales (our simulation posited a single large switch); and would, of course, consider cases where cable TV and telephone provision are mixed. However, the simulation does suggest that two supposed handicaps to entry, namely, that the necessary penetration rate and cost structures are unfavourable, are not critical. It also reveals that much higher labour productivity than BT apparently now achieves is possible.

If it were decided that regional networks could be established, the PTOs' perception of their interests might change. BT would have greater reason to reorganise itself at local level to respond to competition and to press for more efficient working practices, though achievement of these might take some time. Mercury might well take the view in some areas that it would prefer to extend its own network rather than deal with a regional PTO. If its parent company, Cable and Wireless, were to be reluctant to finance the investment, then Mercury might form a joint venture for the purpose. In earlier years, Mercury did consider extending its network into Scotland in association with a local consortium in possession of wayleave rights. If arrangements of this kind were revived in response to potential entry, it would be a satisfactory outcome for users in the regions.

F. MOBILE RADIO

Mobile and fixed telephony markets have been kept separate by a variety of cost and regulatory considerations. New techniques of radio transmission developed for the mobile radio market have potential fixed link applications. These could be the basis for entry into local distribution. In addition, public mobile radio networks already established are well positioned to integrate their mobile and fixed link services.

On the technical level, the higher costs of connecting customers by radio have in the past confined its use in public switched telephone networks to remote locations. Radio transmissions below microwave frequencies, where the signal is suitable also for mobile applications, have been used for fixed links in private networks in the UK until recently. The rapid growth of the mobile radio market has induced the radio regulatory arm of government to re-allocate these frequencies to mobile uses. Private networks have been offered microwave radio frequencies instead. Entry into local distribution would require a further allocation of frequencies, in bands already occupied; the time taken for this to be arranged would be a constraint on speed of entry.

The introduction of cordless telephones, cellular radio and a variety of other mobile radio services opens up the prospect of competition in voice telephony. Technical innovation and increasing scale of operation are reducing the costs of equipment. Cellular radio is proving to be particularly profitable, but the technology is complex. If calls are made only between fixed points, hand sets would be simpler to make and the switches handling the calls would not need to be cellular. For these reasons, radio-based local networks could have significantly lower costs than existing cellular radio networks. Costs of the order of £500 per customer connected, comparable with local cable connection, are feasible and would enable radio networks to be competitive in local distribution.

Existing private mobile radio networks (PMR), currently denied interconnection with the PSTN, could extend their service if more frequencies were made available. Besides protecting BT's revenues, interconnection is limited to conserve spectrum. But this should not inhibit the release of sufficient frequencies. Interconnection could be offered to public networks, such as the new Band III operators, common base stations and trunked systems which make more efficient use of radio frequencies. The best prospects are probably for the adaptation of the new Telepoint technology to use in local telephone networks. Telepoint is a digital cordless telephone in which the handset is owned by the customer and the base station is owned by the network operator. Although being encouraged as a form of mobile radio service, it can quite simply be used to substitute radio for internal wiring in offices and homes and for the last few yards of cable linking customers to public telephone networks.

Entry by mobile radio network operators need not be confined to local distribution. The national cellular radio networks are already able to connect large users by leased lines to their networks (a form of resale). They are not allowed to construct their own fixed links, even trunk links within their network. Racal in particular would welcome the opportunity to construct its own links, or to obtain additional frequencies to adapt its service for local fixed communications. If permitted to do so, it would be well placed to enter the fixed link market, both locally and nationally. The national and regional Band III networks might also be candidates, but it is too soon to judge whether their principal line of business is going to provide a strong base of customers and profits from which to move into fixed links.

G. SATELLITE SERVICES

We consider satellite-delivered services separately, because of their distinctive cost and regulatory characteristics. The analysis tends to corroborate the conclusions already reached concerning the likelihood of domestic competition. We think that it is own account operators, singly or sharing facilities and, possibly, regional networks which would have the most interest. There seems little scope for a major extension of competition in international communications via satellite.

The particular characteristics of satellite circuits are that their costs are independent of distance and the noticeable time lapse as signals travel over the link. In consequence, satellites have been used in the UK almost exclusively for international communications. Domestically, BT has preferred terrestrial methods of transmission, even to remote areas. Mercury has adopted a similar approach. These uses of satellite technology entail direct substitution of one form of terrestrial point-to-point link by satellite. In the last 10 years, specialised satellite services operating in a different mode, point-to-multipoint, have been developed, first for data and more recently for voice.

The duopoly policy recognised that separate consideration would need to be given to these specialised services. The decision to license them was nonetheless delayed until 1988 by concern over possible breaches of the duopoly, in particular, whether the controlling earth station (the 'uplink') should be run by BT or Mercury. If used for data, it would be within the spirit of current policy to permit service providers to run their own uplinks if, as appears to be the case, this is necessary to enable the market to be developed. If used for voice, the possibility of breaching the protection offered by the duopoly needs to be taken into account.

Potential of Satellites for Private Networks

Satellite technology is well suited to providing *point-to-point* and *point-to-multipoint* transmissions as an overlay (or bypass) of public switched voice or data networks. Many such applications have been developed and implemented in North America. These applications typically provide firms with a private network, for which the control centre and space segment are shared among several users. The market has been supplied by specialist companies rather than by established carriers. Satellite technology has well known disadvantages in the provision of switched voice services, arising from the distance signals have to be transmitted to and from the satellite in orbit. In consequence, market entry in specialised services, while providing competition to public data networks, is unlikely to form the basis of a strategy to enter voice telecommunications services.

By the end of the decade, satellite service providers might have an alternative supplier of space segment and uplink in the form of the DBS consortium, British Satellite Broadcasting (BSB), whose satellite facility is adaptable to provide telecommunication services. The nature of the telecommunications licence to be awarded is not yet clear in detail. BSB may wish itself to enter voice telephony. Its announced plans for DBS services indicate that its satellite will have spare capacity, so that entry into telecommunications at marginal cost is an option. It has permission at present only to provide teletext in addition to its broadcasting services. The economics of DBS operation in BSB's principal business of providing satellite distribution of television transmission direct to homes is outside the scope of our analysis. Unlike cable network franchises, DBS has national coverage and hence the opportunity to compete in carrying long-distance traffic.

The 'footprint' of the DBS satellite would also enable international services to be offered, but considerations of 'economic harm' to international satellite organisations mean that permissions are unlikely.

The treaty setting up INTELSAT (see Glossary), to which the UK adheres, requires that any other public international telecommunication service by satellite must not cause it 'economic harm', effectively ruling out direct competition. The 'economic harm' doctrine has held at bay satellite market entry proposals arising in the USA for some years. The regulatory situation now seems to have stabilised around a position in which entry into public international services is not feasible. Entry into international communications by laying undersea cables is not so constrained. Optical fibre cables may also have lower costs per circuit than satellites. Private cable consortia have been formed for the Atlantic and Pacific crossings. Entry by cable now appears more likely, even though securing operating agreements is still difficult.

At the same time, INTELSAT has responded to the prospect of market entry by reducing its charges on high-volume routes, such as across the Atlantic, and broadening the range of services they offer to participating network operators. These actions have reduced the scope for market entry by rival satellite systems, both in switched voice and in other services. For example, INTELSAT has introduced IBS, a small dish satellite service providing digital international private circuits.

Nonetheless, from the point of view of certain would-be entrants into the voice market, satellite delivery certainly represents an option to be considered. The principal considerations are:

o   The costs of a satellite circuit compared to terrestrial transmission depend on traffic density and distribution as well as distance, with satellite being more economic for lower traffic volumes and longer distances. In short, domestic satellite networks could provide a form of long-distance bypass between centres that are more than 50-100 miles apart, depending on the number of points being served.

o   Leasing space segment capacity from INTELSAT or EUTELSAT (see Glossary) presents no difficulties in terms of availability and cost, now or in the next few years. Access to these organisations is via BT or Mercury, which may require regulatory intervention.

o   The cost of the uplink is critical. BT and Mercury do not offer to run uplinks under their tariffs, so they can discriminate against new entrants. If obliged to deal with them, entry is unlikely to occur.

o   In complex networks, the use of satellite circuits in combination with terrestrial introduces additional network management problems. In voice telephony, calls should not be routed twice via satellite (double hop). This technical constraint lessens the attraction of satellites domestically for operators such as BT and Mercury who are committed to using international satellite routes.

These factors suggest that the principal markets are in private networks. Voice traffic would be likely to predominate but data traffic, which potentially offers higher margins, can be mixed in. Whether satellite is preferred to terrestrial technologies will depend in part on the way entry is permitted. A policy of progressive relaxation of restrictions on the use of private circuits and on network management is likely to be adverse to a satellite service. In March 1988, the Director General decided that BT must connect UK users to a satellite service if they specifically request it. The decision was prompted by PanAmSat, an American company proposing to offer private circuit links between North America and Europe. This may lead to a more rapid spread of satellite-based private networks than previously seemed possible.

FOUR

Future Telecoms Policy

Our main theme has been that the major policy decisions taken when BT was privatised have tended to produce less competition than is now possible and desirable, and unless modified will continue to do so. The duopoly has not produced effective competition. Mercury's present strategy offers an insufficient direct challenge to BT, on price, quality of service or range of customers served. Although it has the potential to do much more, it is not likely to become a formidable alternative to BT in the long term without change in policy. The most likely outcome is that Mercury will grow faster than BT but not so fast as to put at risk BT's profits growth. There should be nothing surprising in a duopoly reaching a mutual accommodation in these ways.

BT faces insufficient pressure to reduce its costs. We do not doubt that reducing costs is a principal long-term aim for top management. But managers can achieve sharp changes only if the interests they negotiate with share their perceptions about the company. The efforts of management are ineffective because the external threats are not seen by the rest of BT as real enough to require a change in working practices, or to affect expectations about pay and productivity. This problem has been recognised by the Director General, but within the constraints imposed on him, he can do little to intensify competitive pressures. He can stiffen price controls and the service obligations on BT. But, by tightening regulation, the prospects for intensifying competition are likely to worsen. Reform can come only from fresh political initiative by the Government. In this section, we set out our recommendations.

A. PRICE CONTROL

The Director General has a complex task: price controls in some form must be maintained during the long transition from a 100 per cent BT monopoly to whatever competitive market structure eventually emerges. How should the transition be managed?

There are two alternative strategies. The scope of price controls could be narrowed to focus on the monopoly areas. This would mean the restoration of Professor Littlechild's original *Local Tariff Reduction Scheme*, which regulated rentals and local call charges only. The second strategy would be to use price controls to protect more effectively the customer group most at risk from adverse price changes. For those with single exchange lines and relatively low calling volumes, Mercury is not an effective alternative. Residential subscribers make up the bulk of this group. We prefer the price control approach, mainly because it is important not to prejudge precisely where the boundary between monopoly and competition lies. Appendix 2 describes what we have in mind.

A second element of price control that is necessary for the transition to competition should be a stricter non-discrimination rule than now exists. Mercury has been a more effective competitor in some markets than in others. Future entry will also be partial. To deter BT from making selective tariff responses without undertaking detailed inquiries into costs, it should be required to offer similar terms to all customers, including resellers of its capacity and other service providers. This would severely limit BT's willingness to lower prices to discourage entry, because to do so would reduce profits where entry was not a threat.

B. OWN ACCOUNT OPERATION

The most powerful and immediate way to increase competitive pressures is by liberalisation of own account operation. Such a strategy is blocked, principally, by current rules which effectively disallow own account networks from mixing leased and privately provided transmission capacity. We fear that, in the next five years, the required boldness needed to remove the block is unlikely to materialise. Our pessimism is founded on the following conjunction of factors.

Government departments have been reluctant to accept that much more intensive use of microwave radio frequencies is feasible in Britain, or that private use (which is technically less efficient than public) is desirable. Although proposals to liberalise this area of regulation are under discussion, little in the way of implementation is expected in the near future. Without access to microwave radio, own account operations would be severely handicapped. Only those networks that could be reasonably sure of linking users by optical fibre would be able to enter the market with an offering superior to leased circuits.

A further part of the explanation lies with the declared priorities of telecommunications policy. Mercury has benefited primarily large users: single exchange line customers, both business and residential, have seen few gains from the duopoly. The Government can now be expected to give priority to competitive developments that have a broader constituency and more immediate impact on public services than would releasing the potential of own account operations.

BT is expected strongly to resist the growth of 'own account' liberalisation. It perceives the lack of choice for major users under present regulations as its best method of sustaining profitable business in this sector, at least while its internal efficiency is still low.

Mercury's attitude towards own account development is ambivalent. Like BT, Mercury wishes to prevent the opening up of leased circuits and private provision. For Mercury, however, liberalisation of own account operations is to be preferred to licensing new PTO networks directly, since the prospect of complementary developments should then be greater. In particular, Mercury could expect to gain a significant share of the new leased circuit business. In order to ensure that own account liberalisation does not develop into competitive public networks, Mercury has stated that it should be confined to leased circuits and not be extended to private provision.

We conclude that own account operations will, for some years, be limited to the use of leased circuits. If our pessimism is well founded, competition will be substantially less effective than it might have been. We now consider what might be done in the next five years, on the assumption that freedom for own account operation will be postponed.

Three-pronged Policy Drive

Leaving aside full development of own account operation, three basic lines of attack seem to be available to UK policy:

1. to increase the pay-offs to Mercury so that it develops its network and services more vigorously;
2. to license local and regional PTOs which could extend competition to areas and customers with no effective alternative to BT;
3. to relax licensing restrictions on resale of leased transmission capacity by large users in particular, and so encourage competitive developments by closed user groups and shared networks, but still maintaining the ban on mixing leased and owned capacity.

The substance of, and practical distinction between, each of these options depends on the specific regulatory changes and licensing moves made. We examine what might be done in some detail.

Our analysis of the prospects for entry into voice telephony included the 'Third PTO' option. We do not think policy can rely on it. The 'Third PTO' is the additional national network to which the duopoly policy statement made explicit reference. The reasons for our conclusion are, partly, that there are very few companies able or willing to build a national network (although one or two can be identified) but, mainly, that the prospective profits from competing with BT and Mercury by means of a network which could plausibly meet the required 'national' standard do not appear to be large or certain. Mercury has found that the costs of implementing interconnection with BT and of meeting PTO obligations have seriously hindered its ability to make profits, despite concentrating on large customer business.

C. MERCURY

We indicated that Mercury's strategy has been to seek concentrated traffic primarily from a few large customers. Its appeal has been based on low prices for leased and switched voice services, rather than on superior quality of service. The savings offered are limited by the high price charged for direct connection to Mercury's network, by Mercury's desire to avoid carrying local calls by indirectly connected customers, and by its few direct international connections. Improving Mercury's ability to become a more effective supplier to large customers should be a priority. The possibilities are:

o to increase Mercury's profits from interconnected calls;
o to improve returns from international calls;
o to speed up desirable regulatory changes, such as revision of the national numbering plan.

It would also be possible to ease the burden of PTO obligations on Mercury.

Improving Interconnection

We have already argued that, with respect to the UK, Mercury cannot reasonably hope for much improvement in the generally favourable terms it now receives for the conveyance of calls over the BT telephone system. As BT's tariff rebalancing has continued since the interconnection agreement of 1985, the payments for interconnected calls display some anomalies. However, the appropriate regulatory stance is not to interfere with the commercial terms of interconnection. Mercury has some scope to improve its offers to customers. As Mercury's local switching capacity increases, it should pay Mercury to be competitive on local as well as long-distance calls.

Other elements of interconnection policy should be re-examined. The present arrangements allow BT too much discretion for delay and for ensuring that interconnected calls are inferior in quality. The procedures for making and paying the costs of new connections are also not working well. The uncertainty over the magnitude of the costs of interconnection may be eased by their referral to the committee set up to resolve disputes over interconnection. However, the committee appears to be working very slowly. Difficulties with numbering schemes cannot be resolved without regulatory intervention, probably in the form of the transfer of control to the Director General. This issue affects the position of other PTO entrants.

On international calls, the recent review by the Director General has acknowledged Mercury's difficulties in securing direct overseas connections by relaxing the principles set out in his 1985 Determination. Mercury had argued (consistently with our own analysis) that the 1985 principles did not allow them an adequate margin on calls to countries where Mercury has been unable to secure a direct connection. Where calls are routed by Mercury over BT's switched international circuits, Mercury had to pay BT a sum equivalent to BT's charge for that call less a small discount to reflect the fact that Mercury had carried the call to BT's international exchange. Mercury now claims to have gained a further, more substantial discount. No details of the new Oftel ruling have been made public. We doubt whether the ruling will make a substantial difference to the rate at which Mercury secures international connections.

Improving Mercury's attraction to overseas operators is, given the existing structure of international operating agreements, primarily a matter of increasing Mercury's domestic share. It thus depends on Mercury's vigour in domestic competition. More direct help on international agreements cannot be a priority, since the realistic timetable to shift overseas attitudes must be too long to affect market building issues before the end of the duopoly period.

Given that some market entry will be permitted after November 1990, the main help the Government can give Mercury is the assurance that its prospective profitability will not be put at risk. We assume, therefore, that no entrant will be given the same wide permissions as Mercury has received. In particular, direct international access will be denied.

D. LOCAL AND REGIONAL ENTRY

Our review of entry at local and regional level indicated that broadband cable network operators are likely to remain few in number for some time, because of the slow growth of their television business. This will limit their value as potential entrants into voice telephony. Profit margins would improve if they could deal at arm's length with BT and Mercury. However, the complexities of telecommunication regulation might then represent too high a sunk cost for each franchise to bear on its own; co-operation between cable operators in the provision of voice services should be permitted. Their bargaining position would further improve if they could also co-operate in the building of transmission capacity. For example, in London and other cities where more than one franchise will be granted, cable operators could be allowed to link their networks directly and to switch calls on each other's behalf. Elsewhere, outlying smaller centres of population too small to justify their own franchise (in the range 10,000-100,000 homes) could be added as extensions to cable franchises. Extensions should be for television as well as telephony. For this, as well as extended rights to lay cable, they would require access to microwave radio frequencies.

Such extended local or regional networks could be licensed *de novo*. The principal merit of such regional networks is that they could expect to connect directly a large number of business as well as residential customers. By comparison with broadband cable and other local operators, regional networks would have a wider potential customer base and access to some trunk traffic. Their licences could be written in such a way as to extend the scope of the duopoly approach to competition. By comparison with the Third PTO, regional networks would have a much stronger incentive to co-operate with Mercury over long-distance and international traffic. This way of increasing competition is complementary to the further encouragement of Mercury.

In the simulation of local entry described earlier, a key point is that the potential entrant had much higher labour productivity than BT. The reasons for this lie chiefly in the assumptions concerning manning levels for new technology fitted from scratch, without the incumbent's inherited position, and supposing a concentration on one type of service (voice telephony) alone. If typical of other possible entrants, this suggests that cost differences are likely to be the basis for profitable entry.

New Entrants and BT's Costs

This factor has an important bearing on likely relations between entrants and incumbents. BT would wish to achieve similar levels of cost to those of entrants. Repeated over its network, a substantial improvement would have a large impact on BT's profit. It is essential to recognise the connection between actual entry and the prospective reduction in BT's costs. *Until entry begins to occur, or is immediately in prospect, BT's unit costs will not change much (as indeed they have not over the four years since privatisation).*

It follows that permitting further entry should not damage BT's profits. BT's strategy would logically be to tolerate entry up to the point where gains from cost reductions were matched, at the margin, by losses from the switching of customers to new entrants. How much entry is 'necessary' from BT's point of view? We can only speculate. Much depends on the interpretation of market signals by BT's trade unions. A plausible situation is that once it is clear that Mercury is expanding substantially via collaboration with local networks, a rapid improvement in productivity will occur. In that event, BT's costs will be cut considerably before customers in many areas will have the practical option of buying from new entrants.

BT's main gain from further entry is likely to be that improvements in costs and productivity will penetrate to activities not themselves directly vulnerable to entry. If effects on its costs were all that were at issue in deciding whether to resist entry, BT might well decide not to oppose it. But in practice new networks are most unlikely to confine themselves to voice telephony. They represent a potential challenge across telecom markets. The enhanced rate of change in those markets might well appear threatening to BT.

Local entry in the form in which we have discussed it depends on its being complementary to Mercury. Is collaboration likely to persist after 1990? This mainly depends on Mercury's development of international business. The progress of C&W in getting agreements with foreign PTTs depends, more than anything, on building Mercury's share of international traffic. Local entrants, being independent of BT, can boost Mercury's share substantially. Domestically, Mercury can concentrate on large customers in business districts and welcome local collaborators in the rest of the market. Local networks do not greatly depend on the largest customers; small PABX customers could well account for the largest share of telephone revenues. The degree of overlap in customers can be reduced further by adjusting the payments from local correspondents to Mercury, to encourage greater concentration on smaller customers. However, the BT-Mercury interconnect Determination sets limits on the terms Mercury can offer to local networks.

Mobile Radio

There are other prospects for entry, as we have noted, but these encounter greater regulatory difficulties, and therefore a longer time-scale. The precise scope for entry into fixed network operation by mobile radio companies or others using mobile radio technology cannot be determined at this stage. A capital cost of about £500 per connection we take to be the benchmark for economic entry into local voice telephony. This already appears to be feasible for projected applications of Telepoint technology.

The basis on which mobile and fixed services have been separated to date has been

(a) the perceived need to husband radio frequencies suitable for mobile radio;

(b) the desire to limit the scope of BT's monopoly.

Recent research has confirmed that, in present conditions, mobile radio users have a higher economic value per megahertz of bandwidth than fixed. In consequence, the availability of radio frequencies for fixed services will be co-ordinated with, and generally subordinated to, plans for new or further expanded mobile services. So mobile applications may well be given priority. In areas of the country where there are constraints on mobile radio assignments, chiefly the major conurbations, it is unlikely that radio-based local telephony networks could be established in the next 10 years. Elsewhere, however, there is no technical or economic constraint on making frequencies below 1 GHz available for new types of fixed services.

E. RESALE

Tariff rebalancing by BT since 1981 has eliminated many resale opportunities. The margin between the price of a long-distance call and two local calls on which the 'simple' reseller depends has diminished substantially. However, current levels of call charges indicate that a margin remains for many routes. US experience indicates that entry by resale operators with spare switching capacity can be profitable where such margins exist. Profits were eventually eroded by changes in tariffs by AT&T and local telephone companies. Those resellers that remain in business in the USA have now installed their own long-distance transmission capacity. That is, like own account operations, resale can be a starting point for competitive public networks.

In the UK, where the introduction of new tariffs is not normally delayed by regulatory procedures, response by existing PTOs could be so quick as to deter entry by resellers altogether. Likely tariff responses by BT to the prospect of resale entry would be to reduce the number of 'b' routes, redefine the boundaries between 'b' and 'a' routes and lower the price differential between these two bands. On the other hand, changes in telephone call charges will also affect the attractiveness of Mercury. BT will not lightly upset the present stability between its tariff and Mercury's. A similar difficulty stands in the way of introducing surcharges for leased circuits used by resellers. This would be an attractive option for BT were it not for the fact that it is not in Mercury's interest to make such a discrimination. Oftel are unlikely to oblige Mercury to levy a surcharge.

Whatever the exact response, enough has been said to indicate the vulnerability of simple resale to PTO actions, unless appropriate regulatory action is taken. To call forth large-scale supply from independent resellers, a non-discrimination rule for leased-line charges is necessary.

Network Management

These prospects would also face professional network management firms which might wish to exploit resale of leased transmission capacity. Professional network management would be encouraged if rules governing the use of leased circuits were relaxed generally in conditions where simple resale was not commercially attractive. This would provide business users with additional options for their own account operations but would not amount to a significant competitive challenge to the PTOs.

Network management centred on particular sites, known in the USA as shared tenant services, is already established in Britain on a small scale (including, for example, 90 Long Acre where Mercury has its headquarters). The scope for shared tenant services has been limited by effective confinement to single buildings. As BT and Mercury have noted with their new 'centrex' offerings, large firms with many sites are an important potential client group for network management. Centrex is not, however, itself an adequate response. Taking up centrex effectively precludes users from taking services from independent suppliers and, possibly, from other PTOs. Multisite users do not at present have the realistic choice of professional network management.

The highest priority should be given to removing constraints on efficient design of private networks. The recent Consultative Document issued by Oftel proposes to relax the rules governing the use of leased lines that link private networks in separate ownership (bilateral private circuits).[1] The sharing of leased lines by businesses occupying the same sites is also proposed. These are useful shifts in policy which, if implemented, would remove some of the particularly irksome restrictions. The decisive step would be to remove the restriction on the interconnection of privately constructed fixed links with leased lines within private networks. This would enable private networks to adopt the most efficient configuration (and provide some competition for BT's analogue leased lines).

The obligations imposed on PTOs represent a significant barrier to private networks that might wish to provide a public voice telephony service. For private networks to be able to carry public traffic, liberalisation of interconnection should not be accompanied by extended PTO obligations. It would be possible to develop towards a policy of permitting entry into public telephone service by first allowing sets of users to merge their own traffic. The proposals in the Consultative Document might allow the market in shared tenant services to expand, but do not go far enough. Leaving aside international issues, the principal further relaxation should be to permit the linking of separate buildings in separate ownership, without as at present requiring a common business interest to be demonstrated. If these links were to be formed from leased lines, they would not breach the duopoly; if they were constructed privately, they would. Permitting the use of leased capacity in this manner would also encourage professional network management to provide an extended form of own account operation. Since the new leases are likely to come disproportionately from Mercury, this development would favour Mercury.

Relaxing the 200-metre rule is another step that can be taken during the duopoly period so long as privately provided links are not used to supply a public service. The limit for carrying calls over

---

[1] Director General of Telecommunications, 1989, *op. cit.*

privately provided links could be set at local call charge areas. The further step of incorporating privately constructed capacity in shared use or managed networks must await a decision on the duopoly. The emergence of many enlarged own account groups, or of large managed networks carrying traffic for many users, could be as effective as having a successful Third PTO in the market. However, to complete the development, own account groups would have eventually to get improved interconnection. They would then be as well placed to provide public services as the independent resellers described earlier. These groups would probably be better placed to traverse the regulatory mire lying in the path of all entrants, because they would be important customers of BT in their own right. In short, liberalising own account operations is a practical route to resale freedom.

F. A STRATEGY FOR REFORM

These proposals can be brought together as a strategy for reform. We set out this strategy as a sequence of liberalising changes in regulation in Box 3. The timing of each change is designed to be consistent with commitments already made by the Government and, more importantly in the longer term, to minimise the distortion of business opportunities by regulation. We would urge, for the same reason, that the full strategy should be announced in advance.

Further liberalisation of voice telephony raises the prospect of triangular interconnection arrangements. The Director General must take the initiative to determine these arrangements. If commercial negotiations are allowed to run their course, as happened between BT and Mercury, market entry will be seriously delayed. As the Director General had detailed discussions with the PTOs but has never publicly explained the basis of his 1985 Determination, new entrants would be at a significant disadvantage in any negotiations.

The interconnection issue is further complicated by the fact that some entrants will have PTO obligations and privileges, whilst others will have neither. Which entrants can expect interconnection with BT and with Mercury on preferential terms is unavoidably a matter of Government policy. We think that the linking of preferential terms to PTO obligations should be ended. Interconnection policy should be developed from Oftel's analysis of potential entry. This is the most important analytical task now confronting Oftel.

> **BOX 3**
>
> ## *Summary of Recommended Liberalisation Moves in Voice Telephony*
>
> *Immediately*
> o Relax restrictions on the use of domestic bilateral private circuits under the BSGL.
>
> *At July 1989*
> o License public resale service providers.
> o Relax 200-metre rule in the BSGL to permit interconnection of privately constructed transmission facilities with public and other private networks within local call charge areas.
>
> *Before April 1990*
> o Develop interconnection rules for public networks.
> o Accelerate implementation of new national numbering plan.
>
> *At November 1990*
> o Remove restrictions on provision of voice telephony by local broadband cable networks, including direct interconnection with other such networks.
> o License additional public networks.
> o Remove remaining restrictions on domestic own account operations run under the BSGL.
>
> *After November 1990*
> o Permit resale of international private circuits.

Improving the Regulatory Process

Our criticisms of the way that the regulatory framework set up by the 1984 Telecommunications Act has worked raises the question: Is new legislation required? Nothing we have recommended *requires* new legislation, but there is little doubt that the efficiency of the regulatory process could be greatly improved. We set out some suggestions. Ideally, the statutory duties imposed on the Secretary of State and the Director General should be more pro-competitive in effect. The current duties give priority to ensuring that producers can finance their services and include a rag-bag of interest groups.

There is a good case for redefining the relationship between the Secretary of State and the Director General. The current division in which the Director General has direct control of licence enforcement, and the Secretary of State issues licences, albeit after receipt of the Director General's advice, arose originally from the political imperative to push forward competitive policy in a specific area, and simultaneously to ensure survival of policy initiatives beyond the expected life-span of the Government. The political debate has now progressed to the point where the benefits of competition in telecommunications are no longer in dispute. Debate is about the scope and timing of competition. Since the Director General's task has to shift increasingly towards encouraging entry, it would seem logical to add to our proposed amendments one that gave the Director General power to license. There would of course still have to be safeguards on the exercise of this power. An appropriate mechanism would be for the Secretary of State to retain the right to review the DG's decisions in this area on appeal when the Director General has refused a licence. This would bring telecoms into line with practice in other sectors where regulation is central to entry, such as civil aviation.

But we think that by far the most important potential reform of the 1984 structure involves the generation and use of information. Oftel is handicapped in carrying out its present regulatory strategy by having to bargain for, and not command, requisite information from BT. This arises from Section 53 of the Act, which underpins the DG's power to require information. It sets the standard by which—if it were necessary—the DG could compel the production of information. The standard is one of what could be compulsorily disclosed in civil proceedings. Condition 52·2 of BT's licence substantially further circumscribes the DG's power. It obliges him to make sure that 'no undue burden is imposed on the licensee in procuring or justifying such information', a reasonable enough precaution, but then adds that the

'Licensee is not required to produce or furnish a report which would not normally be available to it unless the Director considers the report essential to enable him to exercise his functions'.[1]

This makes it all too easy for BT to plead 'not available', leaving the DG with little by way of recourse. The abolition of these qualifying phrases would greatly help. The issue is the more important if, as we argue, Oftel were to concentrate on potential entry. The material

---

[1] *Licence granted by The Secretary of State for Trade and Industry to British Telecommunications under Section 7 of the Telecommunications Act 1984*, London: HMSO, 1984 (Revised September 1987), p. 90.

needed to make judgements here is inevitably more speculative and exploratory than for compliance with price controls or for judging 'rebalancing'.

Improving Flow of Information via Oftel

Further steps are required to lessen the asymmetry of information between BT and potential competitors. This could be achieved if Oftel were to collect and publish more data. A clear distinction between what can, and what cannot, legitimately be passed on to potential competitors has to be made.

We suggest that the distinction must turn on the information which would, or would not, be available to all in an effectively competitive industry. This would mean in practice that BT, and for completeness the other PTOs including Mercury, would be required to disclose to the DG facts such as customer expenditures, to be published by him as industry-wide aggregates. Oftel's standard of what would be reasonable to require and disclose could well be set by what is available in the USA. Since there is no reason for Oftel to provide a free service, and because in a competitive industry such information commands a price, it should charge for these services. It would *not*, however, be obliged to reveal the information contained in its own studies of the likely impact of regulatory change on entry. When such changes are decided upon and announced, the reasoning should, of course, be revealed, alongside supporting evidence specific to the change.

Such changes in licence conditions probably could not be agreed with BT, and an amending Bill is therefore required, to embrace the reforms we have suggested.

Appendices

APPENDIX 1

Relationships Between the Regulatory Authorities

The Secretary of State is responsible for licensing service providers and for approving apparatus to be connected to telephone networks. The Director General of Telecommunications is responsible for ensuring that licensees conform to licence terms and conditions, for dealing with complaints and for advising the Secretary of State. The Director General may also exercise functions delegated by the Secretary of State. For example, apparatus approval and the selection of licensees have been delegated.

Both have a general duty under Section 3 of the Telecommunications Act 1984 to ensure that all reasonable demands for telecommunications services are met and that providers of services are able to finance them. In exercising the functions assigned to them, they must act to promote the interests of users, effective competition, research and development, the location of major users in the UK, the use of the UK as a transit point for international communications, and exports.

The Director General's power to enforce licence conditions consists in making an order requiring compliance. He may make a provisional order first. Before making a final order, he must allow time for representations. An order may be challenged in the courts only on the grounds that it is invalid, incorrectly made or *ultra vires*. Once an order is made, any person affected by the continued breach of the licence condition may sue for damages. The order-making procedure has not yet been used.

The Director General can modify general licences by giving notice of the changes, his reasons and allowing time for representations to be made and considered. The Secretary of State can direct him not to make the modification. The Director General can modify licences granted to individual companies only with their consent. If consent is not forthcoming, the Director General may refer the matter to the Monopolies and Mergers Commission

(MMC) for determination. The Secretary of State may direct him not to do so only in the interests of national security or of foreign relations.

The MMC would apply its normal procedures under the Fair Trading Act 1973 in considering a licence amendment reference, except that its considerations would be governed by the duties set out in Section 3 of the 1984 Act and its report would be to the Director General. If the MMC finds in favour of a licence modification, the Director General can make it or a similar amendment. The Secretary of State can direct him not to make the amendment.

The Director General now exercises the functions of the Director General of Fair Trading in respect of telecommunications. He can therefore make a general monopoly reference of a company to the MMC under the Fair Trading Act 1973 or refer a particular course of conduct under the Competition Act 1980.

The Secretary of State can also initiate such references and can refer mergers to the MMC, as was done in respect of BT's take-over of Mitel Corporation in 1985. When such references are made, the MMC would work within the criteria of the 1973 Act. The effect is to lay telecommunications policy open to scrutiny by the MMC.

APPENDIX 2

A Refinement of RP1-X

In this Appendix, we set out an alternative form of price control based upon the consumption of telephone services by the median user. The idea was first aired in the discussions surrounding Professor Littlechild's 1983 report on the regulation of British Telecom's profitability. He did not refer to this particular version of an RPI-X control. Rather his purpose was to compare his 'Local Tariff Reduction' (LTR) proposal with schemes based on other principles (maximum rate of return; an output-related profits levy; a profit ceiling, and zero controls). So the particular merits of the 'median' variant were not discussed.

Littlechild's LTR was estimated to cover 37 per cent of BT's revenue. It was an explicit attempt to confine price controls to the natural monopoly element. However, bargaining between BT and the Government produced a larger basket, of 55 per cent of revenue, including inland trunk calls. This enlargement allowed BT more scope to rebalance its telephone tariffs. BT's past productivity record indicated to the Government an X of −2·5 at the least; much more than −2·5 might jeopardise the proceeds from privatisation. The compromise reached was an X of −3. The same scenario of raising X and widening the basket has now been repeated, with Oftel as the Government's agent. This time, −4·5 is the minimum deemed acceptable.

Selection of the services to be regulated in 1984 was a response to the view that major business customers, by and large, would increasingly benefit from the Mercury threat to BT. The shape of the prediction was correct, but not its speed. Competition has not yet advanced enough to abandon the price control. If future competition makes the impact now expected of it, large customers, almost wholly business customers, will rapidly develop ways of reducing the unit cost of their telephone expenditure. Thus, even more than in 1984, the appropriate focus of concern for remaining BT monopoly power is the small customer, which we define to include much of small business as well as residential customers by specifying a control on the basket represented in the total bill of the single line customer.

The median bill is specified, rather than the mean, because this is in line with political concerns which focus, where they do so explicitly, on persons who are likely to use their telephone less than average. Examples of this concern surface in the 1984 Act where the interests of the elderly and disabled are made eligible for support by local authority subsidy. BT would be required so to conduct its affairs as to make the index of prices, with weights as revealed by the median bill, conform to a given overall increase related to RPI movements. The volume of all services consumed would enter year by year into the calculation; thus it would reflect changes in customers' consumption of telephone services. There would be *some* weight (albeit small for some services) in the basket for virtually every BT output, including, since we must include potential as well as actual customers, connection charges. Within the overall constraint, BT would be free to change its charges as it saw fit, and so 'rebalancing' could proceed in the manner originally conceived, namely, directed by BT's own search for increasing profits.

The definition of the 'median' bill would not be literally that of the actual single customer who happened to be in the exact median position in a given year. Obviously pinning a control on a sample of one bill is unacceptable. To avoid this, sufficient averaging around the median to avoid idiosyncrasy would have to be done. The median bill, so defined, would probably have rather few calls relative to rental charges, and would be an implicit constraint on BT's current ambitions to raise rental charges. But a good case could probably be made out to support a view that basing price controls on the median would promise higher profits (for a given 'X'). First, it explicitly recognises the right of BT to rebalance all its charges as it sees fit, consistent with satisfying a certain set of customers. It allows changing consumption patterns to be anticipated flexibly.

Most importantly, the median control gives greater encouragement to the unbundling of charges for services, because flexibility in meeting the constraint is increased the more prices and quantities there are available.

Nor would the regulator face the impossibly difficult task of splitting BT's assets into those which are used in the 'natural monopoly' area and those which are not. Or, putting the point in terms of what is actually done now, Oftel would escape the logical difficulty of having to apply price controls to 57 per cent of services, while judging 'reasonable' rates of return on BT's whole output.

A change in the scope of price controls implies a change in X. But how would X be changed? Putting it another way, if the price constraint is meant to be binding, what tax on BT's profits is justified to provide the cross-subsidy? The mechanism for doing this would be very much as it is now. Although the process has not been reported, bargains over X are, we believe, conducted with reference to models of BT's future cash flow, and prospective sources and uses of funds, one BT's, the other Oftel's. Within a framework of macro-economic assumptions, they essentially project present policies forward five years, and allow impacts on profits of possible levels of X, alternative baskets, rebalancing, and likely Mercury impacts to be measured. They are not, and do not purport to be, fully worked out regulatory models.

The centrepiece of the bargaining using these models is productivity's effect on future margins. One side naturally thinks more could or should be done to raise this than does the other. The objective is to settle on a figure for X, and the corresponding basket, which will be acceptable politically. On the one hand, this must show an 'improvement' over the last five years for consumers, but on the other must not be so onerous as to lead to a clearly adverse stock market reaction, which would indicate too great a prospective loss of profit. A settlement will be reached which reflects the higher reaches of what BT thinks is possible with respect to unit cost reduction. The difference which our proposed system would make to this is simply that the regulator would not have to go

The names

(signatures including: Patrick Minford, Adam Smith, J. Buchanan, W. H. Hutt, Milton Friedman, F. A. Hayek, Karl Marx, Alan Peacock, A. A. Walters)

The faces

The world's leading economists write for *Economic Affairs*:
Top left, clockwise: **Professors Milton Friedman; F. A. Hayek; Patrick Minford** and **James Buchanan**.

Editor: Robert Miller

Editorial Consultant: Arthur Seldon CBE

Published bi-monthly in association with The Institute of Economic Affairs by City Publications Ltd.

Economic Affairs is the only magazine to offer an international platform for market economists – and their critics – analysing current or impending events and trends in political economy. Each issue has a major theme devoted to a subject of topical interest. Essential reading for economists, students, businessmen, politicians, teachers – and everyone interested in economic affairs.

The magazine

SUBSCRIBE TODAY!

RATES

INDIVIDUAL:
- ☐ UK & Europe £15.00
- ☐ Outside Europe $45.00

INSTITUTION:
- ☐ UK & Europe £27.50
- ☐ Outside Europe $65.00

STUDENT:
- ☐ UK & Europe £10.00
- ☐ Outside Europe $30.00

☐ Please invoice me/my organisation*

NB: Individual rate applies to educational establishments

* delete where inapplicable

Please send me ___ copies of ECONOMIC AFFAIRS for one year (six issues)
I enclose my cheque/P.O. for £/$_____ (Cheques payable to 'Economic Affairs')
NAME (Block Capitals)_____
POSITION (If Applicable)_____
INSTITUTION (If Applicable)_____
ADDRESS (Home/Institution*)_____

POST CODE

☐ Please charge my credit card: £/$_____
Signature_____ Please tick ☐ VISA ☐ Access ☐ American Express ☐ Diners ☐
Date_____ Card No. [ ][ ][ ][ ][ ][ ][ ][ ][ ][ ][ ][ ][ ][ ][ ][ ]

Telephone
0582-32640
with your credit card details.

Please tick where relevant
☐ This is a RENEWAL
☐ This is a NEW SUBSCRIPTION
☐ I would like information on the IEA's PUBLICATIONS' SUBSCRIPTION SERVICE

SEND PAYMENT TO: Economic Affairs, Magazine Subscription Dept., FREEPOST, Luton, LU1 5BR, UK.

RM42

IEA PUBLICATIONS Subscription Service

An annual subscription is the most convenient way to obtain our publications. Every title we produce in all our regular series will be sent to you immediately on publication and without further charge, representing a substantial saving.

Individual subscription rates*

Britain: £25.00 p.a. including postage.
£23.00 p.a. if paid by Banker's Order.
£15.00 p.a. to teachers and students who pay *personally*.

Europe: £25.00 p.a. including postage.

South America: £35.00 p.a. or equivalent.

Other Countries: Rates on application. In most countries subscriptions are handled by local agents. Addresses are available from the IEA.

* These rates are *not* available to companies or to institutions.

To: The Treasurer, Institute of Economic Affairs,
2 Lord North Street, Westminster,
London SW1P 3LB

I should like to subscribe from ....

I enclose a cheque/postal order for:

☐ £25.00

☐ £15.00   I am a teacher/student at ....

☐ Please send a Banker's Order form.

☐ Please send an invoice.

☐ Please charge my credit card:

Please tick ☐ VISA ☐ Access ☐ American Express ☐ Diners

Card No: ☐☐☐☐ ☐☐☐☐ ☐☐☐☐ ☐☐☐☐

In addition I would like to purchase the following previously published titles:

....

Name ....

Address ....
....
....
Post Code ....

BLOCK LETTERS PLEASE

Signed .... Date ....

RM42

through the difficulties of dressing up the outcomes of such a negotiation as consistent with attempts to 'mimic' competition. It is to be hoped that the growth of competition will, by 1993, make it politically feasible to consider abandoning the control. If this does not materialise, the horse-trading will have to be gone through again.

Select Bibliography

K. Baker, *Statement to the Standing Committee on the Telecommunications Bill*, 17 November 1983.

M. E. Beesley, *Liberalisation of the Use of British Telecommunications' Network*, London: HMSO, 1981.

Communications Steering Group, *The Infrastructure for Tomorrow*, London: HMSO, 1988.

CSP International, *Deregulation of the Radio Spectrum in the UK*, London: HMSO, 1987.

Director General of Telecommunications, *Further Deregulation for Business Users of Public Telecommunications Systems*, London: Oftel, February 1989.

Home Office, *Broadcasting in the '90s: Competition, Choice and Quality*, Cm. 517, London: HMSO, 1988.

P. W. Huber, *The Geodesic Network: 1987 Report on Competition in the Telephone Industry*, US Department of Justice, 1987.

S. C. Littlechild, *Regulation of British Telecommunications' Profitability*, London: HMSO, 1983.

Director General of Telecommunications, *The Regulation of British Telecom's Prices*, London: Oftel, January 1988.

Glossary

Access Charges: British terminology differs from American. In this country, access charges can be levied by a network operator with obligations to provide specific loss-making services on other interconnected networks in order to offset these losses. Resale networks, without comparable obligations to those placed on BT and Mercury, might qualify to pay access charges. In the USA, access charges are paid by customers of local telephone companies for the right to make long-distance calls.

Analogue: a form of electromagnetic communication in which the signal frequency varies continuously in line with variations in the pitch of the voice or other message being transmitted.

BSGL: The Branch Systems General Licence governs the way most telephone users connect to public networks. Under its provisions, users may not install their own telephone lines unless confined to a single building or to a single set of premises in single occupation, unless the lines do not extend beyond these limits over an area more than 200 metres across (the 200-metres rule). Branch systems are therefore confined to a single site. Private networks connecting several sites may be operated provided either that each site is linked by means of leased lines or the network is not connected to any other network.

Broadband Cable: A method of transmission by cable capable of carrying television signals.

CCITT: the International Telegraph and Telephone Consultative Committee (the initials refer to the French version of this title) is the organ of the International Telecommunications Union (ITU) whose functions are to study and to make recommendations on technical, operating and tariff questions. Its members comprise the administrations of all ITU member-countries and recognised private operating agencies (RPOAs). BT and C&W are RPOAs.

Cellular Radio: a mobile telephone service that permits users to make and receive calls throughout the country. It is cellular in that radio transmitters serve only a limited area (a cell); as users move from one cell to the next, a central computer automatically hands over calls in progress.

Centrex: a centrex service is provided by public networks as a direct substitute for an internal PABX (*q.v.*) network.

Closed User Group: a regulatory device intended to limit commercial use of telecommunication facilities by restricting permission to members having a common non-telecommunication business interest. Widely used in connection with the shared use of international private circuits, a closed user group is required to be formed for this purpose under the UK VADS (*q.v.*) licence.

Digital: a form of electromagnetic communication in which the signal is composed of binary numbers.

Economies of Scope: the reduction in unit costs obtained when the production of more than one good or service is undertaken simultaneously.

EUTELSAT: the European Telecommunications Satellite Organisation was set up by treaty to operate telecommunications links by satellite on behalf of member countries.

First Instrument Monopoly: a regulation requiring telephone users to rent at least one telephone from the network operator to whose service they are connected.

Fixed Links: a term used to describe permanent communication paths between any two places.

INTELSAT: the International Telecommunications Satellite Organisation was set up by treaty to operate telecommunications links by satellite on behalf of the public telephone services of member countries.

Microwave: radio transmissions at very high frequencies that have the characteristic that the radio signal can be beamed accurately between one radio tower and the next.

Multiplexer: an item of equipment that divides a telephone circuit into several channels.

Network Code of Practice (NCOP): the set of rules devised to ensure that calls routed between branch systems maintain minimum standards of audibility. A provisional NCOP was published by Oftel in December 1986.

OCP (Optional Calling Plan): introduced by AT&T in order to offer a discount to customers making a high number of calls. These customers were perceived to be particularly likely to be attracted by the lower prices charged by AT&T's competitors. BT proposed a similar scheme in 1986 when Mercury commenced switched voice services, but has yet to introduce it.

PABX: a Private Automatic Branch Exchange is the equipment used to route telephone calls over a private network and to and from public networks. A *PABX network* is a private network that is connected to a public network.

Point-to-Point: a communication path that can only connect one place with one other place. A *point to multipoint* communication path connects one place to more than one other place.

Private Network: a telephone network that is not used to provide commercial services.

PSTN (Public switched telephone network): the traditional term for the assets employed in providing commercial telephone services.

PTO: a PTO is the operator of a public telecommunications system. This designation is formally applied by the Secretary of State to telephone networks run under licences that he has granted and which contain obligations to provide particular services, not to show undue discrimination and to trade at published tariffs. In return for these obligations, a PTO would normally, but not invariably, obtain the rights set out in the Telecommunications Code, such as the right to dig up the streets to lay cables. Cable television and cellular radio networks are PTOs. A *Long Line PTO* is authorised to provide telephone service over an area more than 50 km. across (at present, BT and Mercury only). A Long Line PTO has improved rights of interconnection with other networks. PTO is commonly used to mean the operator of a national network, as in the expression 'Third PTO'.

Resale: a generic term covering the commercial use of telephone circuits rented or leased from public networks (leased lines). Resale has traditionally been restricted in order to enable public networks to maintain tariffs that favour particular customers. For example, at the time of privatisation, BT had been favouring business customers by holding down charges for analogue leased lines. '*Simple resale*' is a commercial service in which telephone calls are routed between two local call charge areas over leased lines. By this means, a competitive long-distance telephone service can be provided without investing in additional transmission facilities.

RPI-X (pronounced 'R P I minus X'): The generic term for a price control formula which sets the upper limit of price changes per period as X percentage points below the rate of inflation as measured by the Retail Prices Index.

Tariff Rebalancing: changes in telephone charges intended to align them more closely with costs.

Telepoint: a mobile telephone service which requires the use of a digital cordless telephone.

VANS/VADS: in telecommunications, value added services comprise calls whose content is changed in a definite way in the course of transmission over a telephone network. The 1982 Value Added Network Services (VANS) licence liberalised these services provided that all calls were changed in one or more of the following ways:

(a) being stored for a time;

(b) being altered in format, code or content;

(c) being sent to more than one person.

This definition was quickly seen to be inadequate to cover the variety of data communication services being developed and in the protection of users from the potential abuse of monopoly power. The 1987 Value Added and Data Services (VADS) licence introduced a two-tier licensing structure, in which value-added services are exempt from some conditions concerning trading practices applied to other data services.

Voice Telephony: the telecommunications service for which telephones only are used.